Shark Ecology

Contents

About the Editors

Primo Micarelli

Primo Micarelli, Department of Physical Sciences, Earth and Environment, University of Siena. Head of the Sharks Studies Centre-Scientific Institute (MIPAAF n.1057) of Massa Marittima in Italy and Curator of the Aquarium Mondo Marino. Author or co-author of dozens of international scientific publications. Since 2000, he deals with Biology and Ecology of Elasmobranchs, and Husbandry of Elasmobranchs.

Francesca Romana Reinero

Francesca Romana Reinero, Sharks Studies Centre-Scientific Institute (MIPAAF n.1057) of Massa Marittima, Italy. She is a Marine Scientist and Researcher in Biology, Ethology, Ecology, and Ecotoxicology of sharks. Since 2020, she is the Scientific Coordinator of the Sharks Studies Centre-Scientific Institute of Massa Marittima in Italy. Her research projects focus on Behaviour, Ecology, Biology, and Ecotoxicology of different shark species and she has dozens of publications.

diversity

MDPI

Article

Evidence of Non-Random Social Interactions between Pairs of Bait-Attracted White Sharks in Gansbaai (South Africa)

Primo Micarelli [1,2,*], Francesca Romana Reinero [1], Riccardo D'Agnese [1], Antonio Pacifico [1,3], Gianni Giglio [4] and Emilio Sperone [4]

1 Sharks Studies Center—Scientific Institute, 58024 Massa Marittima, Italy
2 Department of Physical Sciences, Earth and Environment DSFTA, University of Siena, 53100 Siena, Italy
3 Department of Economics and Law, University of Macerata, 62100 Macerata, Italy
4 Department of Biology, Ecology and Earth Sciences DIBEST, University of Calabria, 87027 Rende, Italy
* Correspondence: primo.micarelli@unisi.it

Abstract: Knowledge about the social behavior of sharks is a growing research field, but not many observations are available on the social interactions between pairs of sharks in the presence of passive surface bait and mainly related to aggregations. Between 2009 and 2018, in Gansbaai, South Africa, 415 white sharks were sighted, and 525 surface-generated social interactions were identified, exhibited by 169 different white sharks. The mean sighting rate was 0.91 (range 0.18–1.53) white sharks per hour. Eight patterns of social interaction were exhibited: swim by, parallel swim, follow/give way, follow, give way, stand back, splash fights, and piggyback. Non-random interactions occurred when pairs of specimens approached the passive surface bait, confirming that the white sharks made a real choice, showing a dominance hierarchy during the ten years of data collection. Evidence of non-random social interactions in the surface behavior of bait-attracted white sharks *Carcharodon carcharias* in Gansbaai's transient population was the goal of this research.

Keywords: social behavior; elasmobranchs; behavior; *Carcharodon carcharias*

1. Introduction

Social interactions occur when the behavior of an individual specifically modifies that of another of the same species [1]. Social behavior consists of interactions among conspecifics and results in relationships of variable form, duration, and function. It provides a broad array of behavioral phenomena, including many of the complex forms of cooperation and conflict that are of particular interest to behavioral biologists [2].Social behavior is usually composed of a wide range of modules that affect different contexts of the individual's life, including reproduction, sexual segregation, game behavior, schooling or grouping, cooperation, aggression, and predatory activity; it consists of both genetic bases and precise behavioral units that constitute an intraspecific code that can be instinctive or learned during life [3].

The most observed social modules are those related to aggression [4–7]. Intraspecific competition occurs when two or more organisms of the same species have simultaneous access to a limited resource [8], and the access is commonly established through competitive events that rarely take the form of a direct confrontation. The signalman thus gains an advantage should the recipient decide to withdraw [9]. Intraspecific aggression, in addition to being instinctive, is one of the greatest nuances of sociality: threats, skirmishes, and fights have such a low intensity as to rarely reach harmful consequences for the contenders, thus revealing the presence of any braking mechanisms that promptly interrupt hostilities [3].

Such behaviors can be considered social, as sociability involves several individuals living and/or interacting together, which can lead to the formation of complex social structures [10–13] were the first to conduct research on the social behavior of elasmobranchs, carrying out the first quantitative analyses of domination and subordination in captive specimens of the dogfish *Mustelus canis*. Subsequently, Myrberg and Gruber (1974) [14] identified a dominance hierarchy among scalloped hammerhead sharks *Sphyrna tiburo* depending on size and an apparent predominance of males over females. The direct observations and the passive acoustic tracking of the scalloped hammerhead shark also showed a social behavior that involved the grouping of specimens in a small area, the so-called home range [15].

Social interactions among white sharks, however, have received little attention, but increasing evidence suggests that this species is socially complex [16–19]. Some studies have reported that white sharks often congregate near whale or seal carcasses, and several discrete categories of social interactions have been defined [16,20,21].

As far as white sharks are concerned, the braking mechanism seems to be represented by the establishment of a dimensional hierarchy: the large specimens are dominant compared to the smaller ones, which are identified as subordinate specimens [22,23]. The size, however, is not the only factor influencing the establishment of the hierarchy, but the direction of some relationships of dominance is typically determined by an asymmetry in the competitive behavior of individuals towards each other [24]; in the case of the Gansbaai (South Africa), white shark population, it particularly occurs among specimens of similar sizes. Sperone et al. (2010) [17] suggested that, in the case of sharks of similar lengths, animals need to adopt rituals to establish the social hierarchy: the individuals that manifest all or most aggressive attitudes is considered dominant, while the individual who receives them or initiates most of the acts of submission is considered subordinate. These dominance relationships form the hierarchical structure of the specimens [10,25]. In this study, the long-term observation of the multiple social-interaction patterns exhibited by white sharks with the help of video recordings proved to be useful to evaluate their frequency and priority in establishing the hierarchical relationships between pairs of sharks in the presence of passive prey.

Knowledge about the social behavior of sharks is growing, based on the data that researchers are collecting in different areas of the world. Findlay et al. (2016) [26], in Mosselbay (South Africa), who tried to evaluate whether the white shark forms non-random associations with conspecifics in coastal environments or whether the simultaneous presence of several specimens at the same time and space is entirely due to chance, indicated that the aggregations formed by a very conspicuous number of specimens during certain periods of the year are of an asocial type and due to the presence of external factors, such as the availability of prey, or for reproductive purposes. Subsequently, Schilds et al. (2019) [19] stated that white sharks in the Neptune Islands (South Australia) did not randomly co-occur with their conspecifics but formed four distinct communities or "social groups" in areas where large numbers of white sharks can be sighted throughout the year, including during periods of low seal abundance. Schilds et al. (2019) [19] hypothesized that the observed sex-dependent variations in co-occurrence were linked to intraspecific competition for resources, providing new insights into the aggregatory behavior of white sharks at a seal colony.

Gansbaai is a white shark aggregation site recognized worldwide, and it is the only location in the world where cage-diving trips operate daily, weather permitting [27]. White sharks frequently prey on Cape fur seals (*Arctocephalus pusillus pusillus*) at rookeries off southern Africa and particularly during autumn and winter, when juvenile seals leave to forage offshore for the first time, making them vulnerable to predation by white sharks [28–31]. Gansbaai is a white shark hotspot because the white sharks predate on the resident seals, but the diet also includes fishes and other elasmobranchs. It is an important seasonal feeding ground among individual sharks rather than an adult aggregation site, or a pupping or nursery area [32].

In the present study, we aimed to confirm whether the social interactions that occurred between pairs of white sharks' specimens, when they approached surface bait in the Dyer Island Nature Reserve in Gansbaai, South Africa, were non-random events, and that such interactions were not linked to chance but linked to choice, which aimed at establishing a dominance hierarchy.

2. Materials and Methods

2.1. Study Area

Observations were performed at Dyer Island's Nature Reserve, which is located on the continental shelf 7.5 km off Gansbaai, South Africa (34°41′ S; 19°24′ E) and includes two islands (Figure 1): Dyer Island (the larger) is a low-profile island ca. 1.5 km long and 0.5 km wide, and it is characterized by the presence of different sea bird colonies; Geyser Rock (the smaller) is ca. 0.5 km long and 0.18 km wide, and it hosts a colony of Cape fur seals *Arctocephalus pusillus pusillus* (Schreber, 1775). The reserve is located in the Agulhas Bioregion, which is the meeting point between the Benguela Current, which is the meeting point between the Benguela Current (the eastern boundary current of the subtropical vortex located in the southern Atlantic Ocean) and the Agulhas Current (the current forming the western limit of the Indian Ocean). In summer, intensified south-easterly trade winds result in upwelling, causing the cold waters of Benguela origin to enter the bay (Jury, 1985). The upwelling along the coast results in high biological productivity, which in turn supports large fish stocks, including pilchard, anchovy, and hake [33]. A white shark population is regularly present between March and September in the waters off of the Dyer Island's Nature Reserve. Micarelli et al., 2021b, stated that the low number of resightings recorded in Gansbaai, five in 11 years, corresponding to 1.2% of the population, showed that the white shark population attending this area has a transient behavior. Towner et al. (2013a) [34], in Gansbaai, identified 532 unique individuals, resulting in an estimated super-population size of 908, mainly made up of immature females with a sex ratio of 1:2.2:0.8 for males, females, and unsexed, respectively [32].

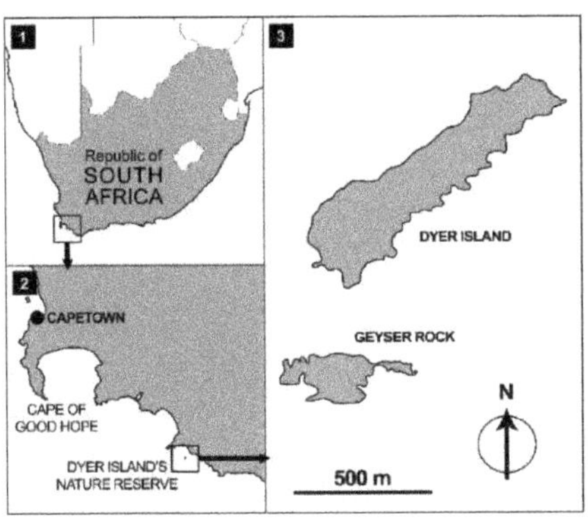

Figure 1. Map of Dyer Island's Nature Reserve in Gansbaai, South Africa (34°41′ S; 19°24′ E).

2.2. Data Collection

During ten scientific expeditions performed between March and May 2009–2018 in the study area, social interactions among white sharks were observed and recorded, and behaviors were later identified from the videos using social displays described by Martin (2003) [16]. White sharks are more abundant around the study area during the austral autumn because they feed on young cape fur seals [17]. During our activities, recorded water temperatures ranged from 13.5° to 18 °C, and underwater visibility from 2 to 5 m (measured to the nearest 0.5 m with a Secchi disc). In total, there were nearly 420 hours of direct observation from the boat, including approximately 200 h from the diving cage. Observations occurred aboard the boat "Barracuda" (Shark Diving Unlimited owner), a 12 m long boat, between 2009 and 2013, and between 2014 and 2018 aboard "Slashfin" (Marine Dynamics), a 14 m long boat. A 4 m long, 3 m high, and 2 m deep rectangular floating cage made of galvanized steel was used for underwater observations; it housed up to three researchers at a time and was moored to the side of the boat. White sharks were sighted at two areas in Gansbaai, Geyser Rock, and Joubertsdam, as reported by Towner et al. (2013a) [34] and Micarelli et al. (2021a) [35], because sharks sighting areas in the Gansbaai gulf changed for unknown reasons, moving from the first area to the second area, and the boat was anchored in both locations at a depth of ~8–10 m, with similar bottom characteristics.

Sharks were attracted to the area around the boat by chumming, following methodologies of the Ferreira and Ferreira (1996) [36], Laroche et al. (2007) [37], and Sperone et al. (2010) [17]. The chum was a mixture of seawater, cod liver oil, fish blood, and pilchards, and an additional 2–3 kg of slices of tuna were used as bait, which were maintained at the sea surface by floats following the methods described by Sperone et al. (2010) [17] and Micarelli et al. (2021a) [35]. Observations from the boat lasted 6–8 h per day, whereas underwater observations lasted 2–4 h per day.

The sex and maturity of sharks were determined by underwater observations from the diving cage. Following Compagno et al. (2005) [38], males with total length (TL) <3.5 m and females with TL <4.5 m were considered immature specimens. Shark length was estimated to the nearest 0.5 m based on observations as sharks passed in front of the measured diving cage of 4.0 m length. Shark sex was determined based on the presence or absence of claspers by observing and filming the pelvic area. Identification of individuals was obtained through photo identification and compilation of specific identification cards, and it was also based on a larger pattern, including not only the different notches of the dorsal fin, but also the following characteristics: caudal fin features, pelvic fin patterns, presence or absence of claspers, gill slashes and body patterns, presence of scars, and/or presence of ectoparasites [32,39]. The identification of the specimens was necessary to verify the dominant and the recessive specimen, as reported in the identification forms compiled by the observers to optimize the analysis during each annual expedition. The surface behavior of white sharks was always recorded by the same team using digital photo cameras and digital video cameras and identified and confirmed from the video using social displays described by Martin (2003) [16].

All observations considered for this paper refer to interactions between two sharks. In our sampling model, aimed at identifying and recording the animals that exhibited social behaviors, a virtual "observation arena" with free-living sharks was drawn to optimize the team's visual sightings from the boat.

This arena, defined as a function of the possibility of observing behavior on the surface up to a maximum depth of ~2–3 m depending on the daily visibility, was bounded in a rectangular area whose longest side was represented by the boat and by ~10 m from the boat's aft and bow, while the shorter side corresponded to ~10 m of distance perpendicular to the boat (Figure 2). The surface area of the "observation arena" was approximately 32×10 m (=320 m^2) between 2007 and 2013 and 34×10 m (=340 m^2) between 2014 and 2018. The bait of floating tuna slices was placed inside the arena, with the aim of activating social behavior (occurring in competitive situations and including aggressive, submissive, and defensive behaviors) of an individual that qualifies, modifies, or otherwise alters the act of another individual of bait-attracted white sharks. As soon as the animals entered the "observation arena" in the direction of the bait, the observations began. An interaction began when two sharks approached each other within a distance of two body lengths, and it ended when two sharks were more than two body lengths apart, moving in different directions, and were not observed together again for at least 2 min, following methods described by Sperone et al. (2010) [17].

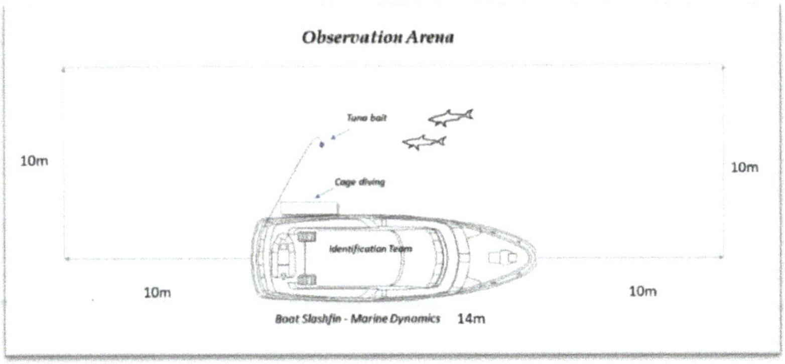

Figure 2. Virtual "observation arena" based on "Slashfin", the 14 m long boat (Marine Dynamics).

2.3. Statistical Analysis

Tests for the linear-independence hypothesis were carried out, through the use of chi-square statistics, to verify the independence between the variable pairs of sex/behavior exhibited, maturity/behavior exhibited, and sex/maturity on behavior exhibited.

To investigate the presence of causality between behavior and bait, we used Cochran's Q test. Cochran's Q test is based on the null hypothesis (H_0), where there is no significant difference in the effectiveness of treatments (the choice is causal) and the alternative (H_1), where there is a significant difference in the effectiveness of treatments (the choice is not causal). It is a non-parametric statistical test used to verify whether k treatments (or number of studies) have similar effects. Generally, the test statistic refers to two-way randomized block design, where the response variable takes only two possible outcomes, coded as 0 or 1, denoting failure or success, respectively. It is Often used to assess whether different observers of the same phenomenon have consistent results (interobserver variability). All statistics were made using Excel 16.44.

3. Results

In total, 415 sharks were sighted, and 525 surface-generated social interactions were identified and exhibited by 169 different white sharks (Figure 3).

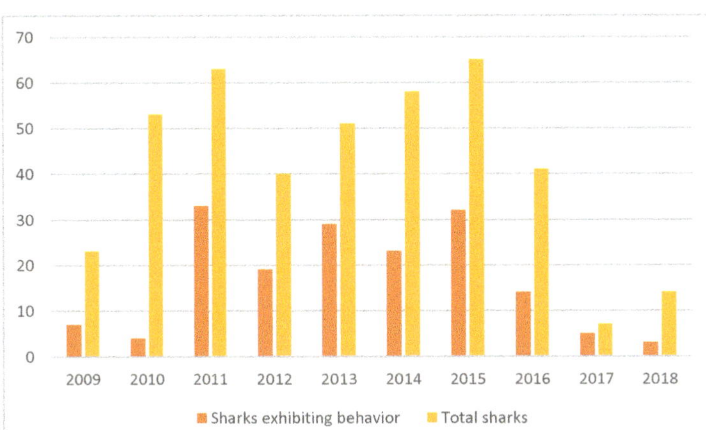

Figure 3. White sharks (*Carcharodon carcharias*) exhibiting social behavior each year and the total number of white sharks sighted per year.

Observed social interactions were classified into one of the following eight interaction patterns as described by Martin in 2003 [16] (Figure 4). Most of the sharks appeared individually in the "observation arena" during the expeditions. The reduced numbers of observations that occurred in 2017 and 2018 were linked to the decline in white shark sightings observed in Gansbaai since 2017 because of the presence of a pair of killer whales (Micarelli et al. 2021b) [35]. The displays are listed in descending order with respect to the performance percentages observed between 2009 and 2018 (Figure 5).

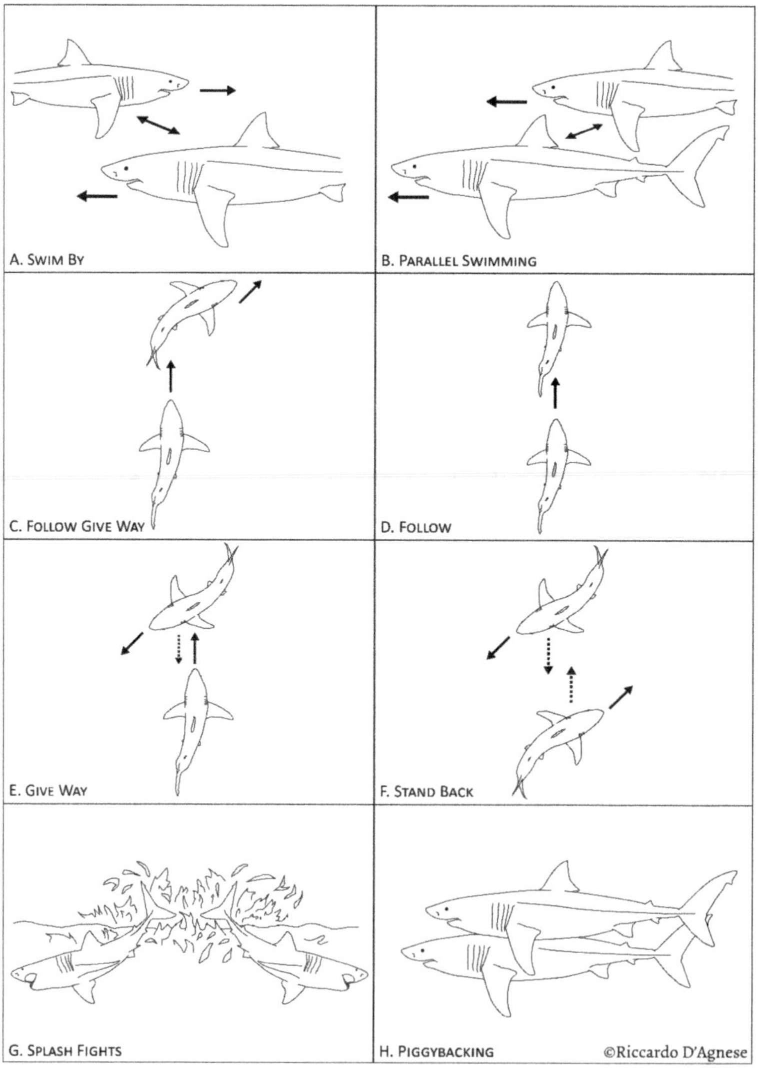

Figure 4. Eight social behaviors observed among white sharks (*Carcharodon carcharias*): swim by (**A**); parallel swim (**B**); follow/give way (**C**); follow (**D**); give way (**E**); stand back (**F**); splash fight (**G**); piggyback (**H**).

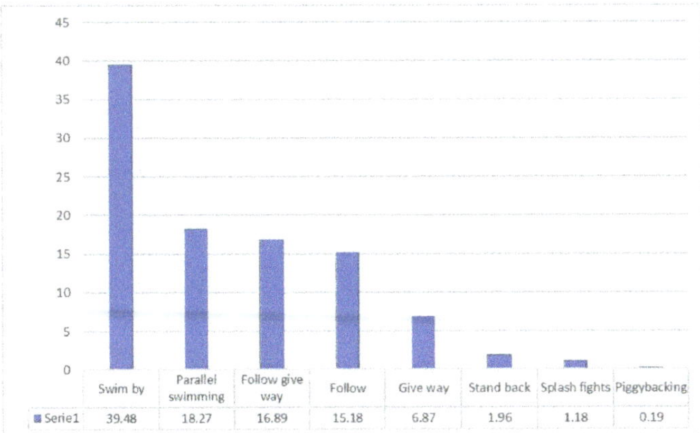

Figure 5. Percentage frequencies of social behavior modules observed between 2009 and 2018 in white sharks (*Carcharodon carcharias*).

Swim by: two sharks in parallel but not on collision courses swam slowly past one another at a close range, usually between 0.5 m and 2.5 m (Figure 4A). The end of the interaction occurred when the sharks passed each other. This display was exhibited during the observations in 39.48% of cases.

Parallel swim: two sharks swam close to one another and maintained a parallel course in the same direction (Figure 4B). The distance between the sharks ranged from 0.5 m to 2.0 m. The interaction usually ended when the followed shark moved away and did not reappear within 5 min. This display was exhibited during the observations in 18.27% of cases.

Follow/give way one shark followed another, causing the followed one to give way. The interaction ended when the followed shark turned left or right at an angle of about 45–90° and did not reappear within 5 min (Figure 4C). The distance between sharks usually ranged from 1 m to 2.5 m. This display was exhibited during the observations in 16.89% of cases.

Follow: one shark followed the other one, repeating its movements and usually with the jaw gaping (Figure 4D). The display ended when the followed shark moved away in a straight direction and did not reappear within 5 min. In this case, the distance between the sharks usually ranged from 0.5 m and 2 m. This display was exhibited during the observations in 15.18% of cases. (*The differences between follow/give way and follow consist in the fact that in the first case, the followed shark quickly moves away with a sharp angle of 45–90°, while in the second case, the following shark keeps its mouth open and repeats the movements of the followed one and this moves away fast without veering*).

Give way: two sharks swam on a collision course, and the interaction ended when one shark turned left or right at an angle of ~45° and did not reappear within 5 min (Figure 4E). This display was exhibited during the observations in 6.87% of cases.

Stand back: simultaneous deviation of two colliding individuals with no established dominant (Figure 4F). This display was exhibited during the observations in 1.96% of cases.

Splash fight: this interaction began when one shark rolled onto its side at the surface and directed splashes towards the other one; the other could respond by returning similar splashes (Figure 4G). It ended when one of the sharks retreated. This display was exhibited during the observations in 1.18% of cases.

Piggyback: for this behavior, one shark descended onto the back of the other one, and the two animals swam in unison for several seconds (Figure 4H). It ended when the sharks separated. This interaction occurred only once, for 22 s, and between a male and a female shark of similar sizes. This display was exhibited during the observations in 0.19% of cases.

The Chi-square test of independence is a statistical hypothesis test used to determine whether two variables are likely to be related (dependency) or not (independency). This test is very useful when studying variables evaluated through frequency counts. More precisely, the statistic test is obtained by solving for the ratio between the mean squared error from the theoretical frequencies and their weights:

$$\chi^2 = \frac{\sum_i \sum_j \left(n_{(i,j)} - \hat{n}_{(i,j)}\right)^2}{\hat{n}_{(i,j)}} \tag{1}$$

Here, $n_{(i,j)}$ denote the absolute joint frequency, where i and j are numerical indices (raw data) representing two random discrete variables and $\hat{n}_{(i,j)}$ stands for the absolute joint frequency in case of independence. This latter, in the denominator, refers to the weights negatively related with the $n_{(i,j)}$'s significance (dependency). Thus, in the case of a strongly causal (and then estimable) relationship between two variables, $\hat{n}_{(i,j)}$ would decrease, improving the significance of the test (a higher χ^2's test statistic).

In this study, we tested the independence between the following two pairs of variables: exhibited behavior/sex and exhibited behavior/maturity. As highlighted in Table 1, every variable (sex and maturity) shows p-values close to zero, leading to the rejection of the null hypothesis in favor of the alternative of dependency. According to the χ^2's test statistics addressed for every variable accounted for, the highest result is associated with females/immatures. Thus, immature female specimens would have significantly exhibited more social behaviors in 10 years of data collection (Table 1).

Table 1. Chi-square test. Table on the left refers to the test addressing the two pairs of variables. The χ^2's test statistics are 78.43 (Sex); 87.74 (Maturity); 63.69 (Sex/Maturity). The table on the right refers to the test addressed for every variable accounted for. The χ^2's test statistics are 63.21 (Males/Adults); 65.17 (Males/Immatures); 73.61 (Females/Adults); 84.19 (Females/Immatures). The significance codes stand for: *** significance at 1%, ** significance at 5%, * significance at 10%.

Sex/Maturity (X)	**Patterns/Behavior (Y)**
Sex	p-Value 0.00
Maturity	p-Value 0.00
Sex/Maturity	p-Value 0.00
Chi–Square on social behavior (p-Value)	
Total specimen	0.00 ***
Males	0.00 ***
Females	0.00 ***
Adults	0.00 ***
Immatures	0.00 ***

In this analysis, the random (casuality) or non-random (non-casuality) occurrence of the social interactions was tested using Cochran's Q test, where the treatments denote how the presence of the bait affects the sharks' behaviors. It is a non-parametric test applied to the analysis of two-way categorical variables (success = 1, unsuccess = 0). Thus, in technical terms, the test requires a binary variable of interest and assesses whether the proportion of causal links ("successes") is the same between the groups.

In this study, the nominal outcome is a causal and non-causal link according to shark's behavior, where the groups refer to the numerical indices i and j (raw data) of the nominal variable $y_{(i,j)}$. In this way, one would be able to test whether the effectiveness of the treatments is different between the groups. More precisely, we can investigate the presence of heterogeneity between observations without incurring linear problems such as multicollinearity, heteroskedasticity, asymptotic distributions, and exogeneity.

Let the degrees of freedom be $k - 1 = 1$; the resulting p-value is lower than the significance level ($\alpha = 5\%$), so one should reject the null hypothesis in favor of the alternative (Table 2):

Table 2. Cochran's test results: the "Test Statistic" refers to the Chi-squared distribution, the "critical value" is the threshold above which one should reject the null hypothesis, the "df" stands for the degrees of freedom, "p-value" denotes the probability of wrongly choosing the true hypothesis, and "significance level" denotes the probability of rejecting the null hypothesis when it is true (Type I Error).

Test Statistic	175,98
Chi-squared distribution (critical value)	3,840
df	1
p-Value (one-tailed)	0,027
significance level	0,050

***H_0*.** *There is no significant difference in the effectiveness of treatments (the choice is casual).*

***H_1*.** *There is a significant difference in the effectiveness of treatments (the choice is not casual).*

4. Discussion

Not many observations are available on the social behaviors of sharks, particularly about white sharks, and they are mainly related to aggregations. In fact, although sharks are often seen as solitary predators, it has been observed that some shark species can aggregate not simply in response to environmental changes, but in actual groups of mutually attracted individuals [40–42].

Such behaviors have been observed in the following orders: Heterodontiformes [43], Hexanchiformes [44], Squatiniformes [45], and Carcharhiniformes [46,47]. At the same time, evidence of social interactions has also been reported in planktophagouses [48], in large predators [49], in coastal water species [46,50], and in reef sharks [47,51–53]. Several theories have been proposed for the reasons for these aggregations: the formation of groups could be attributed to communication, to the transfer of social information [46], to predation, to the protection of the group [52,54], and to reproduction [55–59].

In addition, the white shark, although mostly solitary, has been reported to aggregate at a number of sites worldwide: these include Seal Island, Gansbaai, and Mossel Bay in South Africa [60], Guadalupe Island in Mexico [61], the Chatham Islands in New Zealand [62], Cape Cod in Massachusetts, USA [63], and Neptune Islands in Australia [64]. Most aggregation sites have been suggested to provide important feeding grounds for white sharks [65], and aggregations have been shown to be driven by the seasonal availability and abundance of prey species [64,66,67]. At the same time, sharks have been observed to develop and maintain complex social behaviors such as dominance hierarchies [13,14] and stable social bonds [68], as well as to be able to learn social information [69,70]; these abilities are due to the fact that sharks are characterized by a high ratio between brain mass and body mass [71–73], comparable to that of mammals.

However, the highly mobile nature of sharks, combined with the difficulty of tracking their movements in the open sea, makes it difficult to study their social interactions.

The hypotheses of intraspecific associations and groupings are mainly based on direct observations in the field [52] and on some recent tracking data thanks to acoustic satellite telemetry [50] or the use of passive receivers [74–76]. By aggregating, individuals facilitate social interactions and mating [77].

Furthermore, some observations have revealed unusual adaptation mechanisms, particularly in some shark species [78]; among these, the seven-gill sharks *Notorynchus cepedianus* seem to use multiple feeding strategies depending on both size and type of prey, adopting strategies of social facilitation and hunting in packs to deal with larger prey such as Cape fur seals *Arctocephalus pusillus pusillus* [44].

The scientific literature has also begun to outline what the traits of the sociobiological background are of the elasmobranchs, and how this comes to structure itself beyond instinct [57,79,80]. Clues to this presumed influence are provided by the different forms of learning currently proven in elasmobranchs, such as those inherent in hunting: individuals can improve the techniques of searching and capturing prey over time, thus increasing their successes [79] and individual surface behavior of white sharks in the presence of bait is not a simple stimulus–response reflex, but rather a complex tactical situation with plastic responses [39].

Although for years the white shark has been considered a solitary species, some studies have led researchers to affirm that, in reality, it is a socially complex species [16,17]. For example [81], in southern California, hypothesized that juvenile white sharks form distinct communities during critical early phases of ontogeny and how a tendency to co-occur across life stages may originate from the formation of these communities in early ontogeny. Schilds et al. (2019) [19], in Australia in the Neptune Islands, found that white sharks did not randomly co-occur with their conspecifics, but formed four distinct communities. Off Guadalupe Island, Mexico, animal-borne telemetry receivers revealed that white sharks varied in the number of associations they formed and occurred most often when white sharks were swimming in straight paths or when they were turning frequently. While many associations were likely random, there was evidence of some stronger associations [82].

Klimley (2000) [83] defined the white shark as an "eavesdropper" species that is capable of interpreting and exploiting the successes of others during foraging; this feature is also called "social learning" or the possibility for some animals to extrapolate information gained from other individuals [79,84]. If it is true that juvenile white sharks form distinct communities, as claimed by Anderson et al. (2021) [81], the establishment of a hierarchy, in the presence of passive surface bait, is favorable for both individuals concerned, as it averts a direct confrontation and, therefore, the possibility of injury, presenting a gain both in terms of energy and the possibility of preserving themselves for future predation [17]. An example of the presence of a hierarchy in white sharks occurs during their scavenging activity, in which two possibilities of behavior are highlighted: either there is an approach to a carcass depending on the size order of the sharks present [23] or, in the event of numerous specimens intent on feeding, the larger ones have access to the portions with more fat and therefore energetically more beneficial, while the subordinate specimens feed on the remainder or pieces floating in the sea. This mechanism allows the reduction in collisions and life-threatening injuries. If there is no competition for a resource, aggression is minimal [85].

The analysis of the data collected in 10 years of observations showed how sharks did not form random social interactions and that this is not linked to chance but linked to choice to create a dominance hierarchy between them, and immature females were more inclined to do so. In Gansbaai, the white shark population is numerically dominated by females [32], probably explaining why females exhibit such behaviors.

Thus, despite having been regarded as a solitary species for many years [38], and if it is true that intraspecific competition ethogram occurs when two or more organisms of the same species have simultaneous access to a limited resource [8] and the access is commonly established through competitive events that rarely take the form of a direct confrontation, the white shark turns out to be an animal capable of exhibiting social behaviors in particular situations.

Specifically, the results of the social interactions exhibited by pairs of bait-attracted white sharks in the Dyer Island's Nature Reserve show both competitive modules of follow and follow/give way, and "mild" observation modules [17], such as parallel swim and swim by: the latter, in fact, as shown in Figure 5, was the most observed social behavior. As already noted in the study by Sperone et al. (2012) [40] in Gansbaai, these "mild" observation modules reflect a type of extremely calm approach and an initial hesitation, thus confirming the studies by Bromilow (2014) [86] and the observations made by Martin et al. (2005) [29], who argued that the white shark certainly has a more aggressive approach with natural prey rather than with passive ones, when the investigation period is of longer duration.

In particular, although this species was considered by many to be myopic [16,87,88], recently, Micarelli et al. (2021a) [36], in the Gansbaai population, stated that white sharks implement their predatory choices to energetically richer prey, thanks especially to their visual ability, which plays an important role in adults and immatures with dietary shifts in their feeding patterns and, as noted in juvenile white sharks in the study by Lisney et al. (2007) [89], the olfactory bulbs are relatively smaller and the optic roof relatively larger than in adults; for this reason, in Gansbaai, immatures tended to be more observant in the hunting area and confronted each other in capturing prey, exhibiting the social interactions with the modules described by Martin in 2003.

In the future, having highlighted and confirmed the presence of non-random social behaviors between pairs of sharks in Gansbaai in the presence of passive surface bait finalized to establish a dominance hierarchy, it would be interesting to apply network analysis methodologies to evaluate the possible presence in this area of distinct social groups, as already observed in the Neptune Islands in Australia [19], or in southern California [82], where non-random co-occurrence of juvenile white sharks at seasonal aggregation sites were recorded. Because Reinero et al. 2022 [90] stated that environmental factors influence the prey discrimination of white sharks in Gansbaai during their surface passive prey predatory behavior and that tide range is the most important factor that influences the white sharks' prey choice, followed by underwater visibility, water temperature, and sea conditions, it will also be important to deepen the understanding of whether social behaviors of white sharks may also be affected by potential factors such as abiotic conditions, natural prey and competitor species, and interspecific interactions.

Author Contributions: P.M., conceptualization, formal analysis, investigation, methodology, writing—original draft and editing; F.R.R., formal analysis, investigation and writing—review; R.D., formal analysis and writing—review; G.G., investigation and writing—review E.S., conceptualization, formal analysis, investigation, methodology and writing—review; A.P., formal analysis and writing—review. All authors have read and agreed to the published version of the manuscript.

Funding: This research received no external funding.

Institutional Review Board Statement: The ethical review and approval were waived for this study because it did not intervene in the observed animals.

Data Availability Statement: Data for this project are maintained by the Sharks Studies Centre—Scientific Institute, the University of Calabria, and the University of Siena, Italy. The data are available from the authors.

Acknowledgments: We are grateful to the CSS team members that carried out 11 expeditions for their indirect financial support of this research, and thanks are also due to Marine Dynamics (Hennie Otto, Wilfred Chivel) for their logistical assistance and all other field assistance with data collection.

Conflicts of Interest: The authors declare no conflict of interest.

References

1. Baldaccini, N.E.; Mainardi, D.; Papi, F. "Introduzione Alla Etologia" Book—by Editoriale Grasso, Bologna. 1990. Available online: https://elearning.unite.it/pluginfile.php/247583/mod_resource/content/1/Testo_Didattica.pdf (accessed on 1 March 2023).
2. Nowak, M.A. Five rules for the evolution of cooperation. *Science* **2006**, *314*, 1560–1563. [CrossRef] [PubMed]
3. Lorenz, K. *On Aggression*; Psychology Press: London, UK, 1966.
4. Silber, G.K. The relationship of social vocalizations to surface behaviour and aggression in the Hawaiian humpback whale (*Megaptera novaeangliae*). *Can. J. Zool.* **1986**, *64*, 2075–2080. [CrossRef]

5. Higley, J.D.; King, S.T., Jr.; Hasert, M.F.; Champoux, M.; Suomi, S.J.; Linnoila, M. Stability of interindividual differences in serotonin function and its relationship to severe aggression and competent social behavior in rhesus macaque females. *Neuropsychopharmacology* **1996**, *14*, 67–76. [CrossRef] [PubMed]
6. Connor, R.C.; Mann, J.; Tyack, P.L.; Whitehead, H. Social evolution in toothed whales. *Trends Ecol. Evol.* **1998**, *13*, 228–232. [CrossRef] [PubMed]
7. Oliveira, R.F.; Simões, J.M.; Teles, M.C.; Oliveira, C.R.; Becker, J.D.; Lopes, J.S. Assessment of fight outcome is needed to activate socially driven transcriptional changes in the zebrafish brain. *Proc. Natl. Acad. Sci. USA* **2016**, *113*, E654–E661. [CrossRef] [PubMed]
8. Wilson, D.S. A theory of group selection. *Proc. Natl. Acad. Sci. USA* **1975**, *72*, 143–146. [CrossRef]
9. Burghardt, G.M. Defining "communication". *Commun. Chem. Signals* **1970**, *1*, 5–18.
10. Hinde, R.A. Interactions, relationships and social structure. *Man* **1976**, *11*, 1–17. [CrossRef]
11. Krause, J.; Ruxton, G. *Living in Groups*; Oxford University Press: Oxford, UK, 2002.
12. Wey, T.; Blumstein, D.T.; Shen, W.; Jordan, F. Social network analysis of animal behaviour: A promising tool for the study of sociality. *Anim. Behav.* **2008**, *75*, 33344. [CrossRef]
13. Allee, W.C.; Dickinson, J.C. Dominance and subordination in the smooth dogfish, *Mustelus canis* (Mitchell). *Physiol. Zool.* **1954**, *27*, 356–364. [CrossRef]
14. Myrberg, A.A., Jr.; Gruber, S.H. The behaviour of the bonnethead shark, *Sphyrna tiburo*. *Copeia* **1974**, *1974*, 358–374. [CrossRef]
15. Klimley, A.P. Social Organization of Schools of the Scalloped Hammerhead Shark, *Sphyrna lewini* (Griffith and Smith), in the Gulf of California. Scripps Institution of Oceanography Technical Report, San Diego, CA, USA. 1983. Available online: https://escholarship.org/uc/item/2qg6s9t5 (accessed on 1 March 2023).
16. Martin, R.A. *Field Guide to the Great White Shark*; ReefQuest Centre for Shark Research: Vancouver, BC, USA, 2003; 185p.
17. Sperone, E.; Micarelli, P.; Andreotti, S.; Spinetti, S.; Andreani, A.; Serena, F.; Brunelli, E.; Tripepi, S. Social interactions among bait attracted white sharks at Dyer Island (South Africa). *Mar. Bio Res.* **2010**, *6*, 408–414. [CrossRef]
18. Becerril-García, E.E.; Hoyos-Padilla, E.M.; Micarelli, P.; Galván-Magaña, F.; Sperone, E. The surface behaviour of white sharks during ecotourism: A baseline for monitoring this threatened species around Guadalupe Island, Mexico. *Aquat. Conserv. Mar. Freshw. Ecosyst.* **2019**, *29*, 773–782. [CrossRef]
19. Schilds, A.; Mourier, J.; Huveneers, C.; Nazimi, L.; Fox, A.; Leu, S.T. Evidence for non-random co-occurrences in a white shark aggregation. *Behav. Ecol. Sociobiol.* **2019**, *73*, 138. [CrossRef]
20. Compagno, L. *Sharks of the World: An Annotated and Illustrated Catalogue of Shark Species Known to Date. Vol. 2 Bullhead, Mackerel, and Carpet Sharks (Heterodontiformes, Lamniformes and Orectolobiformes)*; FAO Species Catalogue for Fishery Purposes, no. 1; FAO: Rome, Italic, 2001; Volume 2.
21. Klimley, A.P.; Le Bouef, B.J.; Cantara, K.M.; Richert, J.E.; Davis, S.F.; Van Sommeran, S.; Kelly, J.T. The hunting strategy of white sharks (*Carcharodon carcharias*) near a seal colony. *Mar. Biol.* **2001**, *138*, 617–636. [CrossRef]
22. Klimley, A.P.; Anderson, S.D. Residency Patterns of White Sharks at the South Farallon Islands, California. In *Great White Sharks. The Biology of Carcharodon carcharias*; Klimley, A.P., Ainley, D., Eds.; Academic Press: Cambridge, MA, USA, 1996.
23. Long, D.J.; Jones, R.E. White shark predation and scavenging on cetaceans in the Eastern North Pacific Ocean. In *Great White Sharks. The Biology of Carcharodon carcharias*; Klimley, A.P., Ainley, D., Eds.; Academic Press: Cambridge, MA, USA, 1996.
24. Chase, I.D. Cooperative and noncooperative behavior in animals. *Am. Nat.* **1980**, *115*, 827–857. [CrossRef]
25. De Vries, H.; Stevens, J.M.G.; Vervaecke, H. Measuring and testing the steepness of dominance hierarchies. *Anim. Behav.* **2006**, *71*, 585–592. [CrossRef]
26. Findlay, R.; Gennari, E.; Cantor, M.; Tittensor, D.P. How solitary are white sharks: Social interactions or just spatial proximity? *Behav. Ecol. Sociobiol.* **2016**, *70*, 1735–1744. [CrossRef]
27. Towner, A.V. Great White Sharks *Carcharodon carcharias* in Gansbaai, South Africa: Environmental Influences and Changes over Time, 2007–2011. Master's Thesis, University of Cape Town (UCT), Cape Town, South Africa, 2012.
28. Martin, R.A.; Hammerschlag, N.; Collier, R.S.; Fallows, C. Predatory behaviour of white sharks (*Carcharodon carcharias*) at Seal Island, South Africa. *J. Mar. Biol. Assoc. UK* **2005**, *85*, 1121–1135. [CrossRef]
29. Hammerschlag, N.; Martin, R.A.; Fallows, C. Effects of environmental conditions on predator-prey interactions between white sharks (*Carcharodon carcharias*) and Cape fur seals (*Arctocephalus pusillus pusillus*) at Seal Island, South Africa. *Environ. Biol. Fishes* **2006**, *76*, 341–350. [CrossRef]
30. Kirkman, S.P.; Oosthuizen, W.H.; Meÿer, M.A. The seal population of Seal Island, False Bay. In *Finding a Balance: White Shark Conservation and Recreational Safety in the Inshore Waters of Cape Town, South Africa*; Nel, D.C., Peschak, T.P., Eds.; Proceedings of a Specialist Workshop. WWF South Africa Report Series–2006/Marine/001 Annexure 1; YUMPU: Diepoldsau, Switzerland, 2006; pp. 83–94.
31. Towner, A.V.; Underhill, L.G.; Jewell, O.J.; Smale, M.J. Environmental influences on the abundance and sexual composition of white sharks Carcharodon carcharias in Gansbaai, South Africa. *PLoS ONE* **2013**, *8*, e71197. [CrossRef] [PubMed]
32. Micarelli, P.; Bonsignori, D.; Compagno, L.J.V.; Pacifico, A.; Romano, C.; Reinero, F.R. Analysis of sightings of white sharks in Gansbaai (South Africa). *Eur. Zool. J.* **2021**, *88*, 363–374. [CrossRef]
33. Griffiths, C.L.; Robinson, T.B.; Lange, L.; Mead, A. Marine biodiversity in South Africa: An evaluation of current states of knowledge. *PLoS ONE* **2010**, *5*, e12008. [CrossRef] [PubMed]

34. Towner, A.V.; Wcisel, M.A.; Reisinger, R.R.; Edwards, D.; Jewell, O.J.D. Gauging the threat: The first population estimate for white sharks in South Africa using photo identification and automated software. *PLoS ONE* **2013**, *8*, e66035. [CrossRef]
35. Micarelli, P.; Chieppa, F.; Pacifico, A.; Rabboni, E.; Reinero, F.R. Passive Prey Discrimination in Surface Predatory Behaviour of Bait-Attracted White Sharks from Gansbaai, South Africa. *Animals* **2021**, *11*, 2583. [CrossRef] [PubMed]
36. Ferreira, C.A.; Ferreira, T.P. Population dynamics of white sharks in South Africa. In *Great White Sharks: The Biology of Carcharodon carcharias*; Klimley, A.P., Ainley, D.G., Eds.; Academic Press: San Diego, CA, USA, 1996; pp. 381–391.
37. Laroche, R.K.; Kock, A.A.; Dill, L.M.; Oosthuizen, W.H. Effects of provisioning ecotourism activity on the behaviour of white sharks *Carcharodon carcharias*. *Mar. Ecol. Prog. Ser.* **2007**, *338*, 199–209. [CrossRef]
38. Compagno, L.; Dando, M.; Fowler, S. *Sharks of the World*; HarperCollins Publishers Ltd.: London, UK, 2005; 368p.
39. Sperone, E.; Micarelli, P.; Andreotti, S.; Brandmayr, P.; Bernabò, I.; Brunelli, E.; Tripepi, S. Surface behaviour of bait-attracted white sharks at Dyer Island (South Africa). *Mar. Biol. Res.* **2012**, *8*, 982–991. [CrossRef]
40. Springer, S. Social organization of shark populations. In *Sharks, Skates and Rays*; Gilbert, P.W., Mathewson, R.F., Rall, D.P., Eds.; Johns Hopkins Press: Baltimore, MA, USA, 1967; 644p.
41. Jacoby, D.M.P.; Croft, D.P.; Sims, D.W. Social behaviour in sharks and rays: Analysis, patterns and implications for conservation. *Fish Fish.* **2011**, *13*, 399–417. [CrossRef]
42. Micarelli, P.; Pieraccini, F.; Reinero, F.R.; Sperone, E. Influence of male presence on the social structure of lesser spotted dogfish (*Scyliorhinus canicula*) female groups. *Int. J. Oceanogr. Aquac.* **2020**, 4. [CrossRef]
43. Powter, D.M.; Gladstone, W. The reproductive biology and ecology of the Port Jackson shark *Heterodontus portusjacksoni* in the coastal waters of eastern Australia. *J. Fish Biol.* **2008**, *72*, 2615–2633. [CrossRef]
44. Ebert, D.A. Observations on the predatory behaviour of the sevengill shark *Notorynchus cepedianus*. *S. Afr. J. Mar. Sci.* **1991**, *11*, 455–465. [CrossRef]
45. Standora, E.A.; Nelson, D.R. A telemetric study of the behavior of free-swimming angel sharks *Squatina californica*. *Bull. South. Calif. Acad. Sci.* **1977**, *76*, 193–201.
46. Klimley, A.P.; Nelson, D.R. Schooling of the scalloped hammerhead sharks, *Sphyrna lewini*, in the Gulf of California. *Fish. Bull.* **1981**, *79*, 356–360.
47. McKibben, J.N.; Nelson, D.R. Patterns of movement and grouping of gray reef sharks, *Carcharhinus amblyrhynchos*, at Enewetak, Marshall Islands. *Bull. Mar. Sci.* **1986**, *38*, 89–110.
48. Meekan, M.G.; Bradshaw, C.J.A.; Press, M.; Mc Lean, C.; Richards, A.; Quasnichka, S.; Taylor, J. Population size and structure of whale sharks *Rhincodon typus* at Ningaloo Reef, Western Australia. *Mar. Ecol. Prog. Ser.* **2006**, *319*, 275–285. [CrossRef]
49. Domeier, M.L.; Nasby-Lucas, N. Annual re-sightings of photographically identified white sharks (*Carcharodon carcharias*) at an eastern Pacific aggregation site (Guadalupe Island, Mexico). *Mar. Biol.* **2007**, *150*, 977–984. [CrossRef]
50. Heupel, M.R.; Simpfendorfer, C.A. Quantitative analysis of aggregation behavior in juvenile blacktip sharks. *Mar. Biol.* **2005**, *147*, 1239–1249. [CrossRef]
51. Stevens, J.D. Biological observations on sharks caught by sport fisherman of New South Wales. *Mar. Freshw. Res.* **1984**, *35*, 573–590. [CrossRef]
52. Economakis, A.E.; Lobel, P.S. Aggregation behavior of the grey reef shark, *Carcharhinus amblyrhynchos*, at Johnson Atoll, Central Pacific Ocean. *Environ. Biol. Fishes* **1998**, *51*, 129–139. [CrossRef]
53. Speed, C.W.; Meekan, M.G.; Field, I.C.; Mcmahon, C.R.; Stevens, J.D.; Mcgregor, F.; Huveneers, C.; Berger, Y.; Bradshaw, C.J.A. Spatial and temporal movement patterns of a multi-species coastal reef shark aggregation. *Mar. Ecol. Prog. Ser.* **2011**, *429*, 261–275. [CrossRef]
54. Wearmouth, V.J.; Sims, D.W. Sexual segregation in marine fish, reptiles, birds and mammals: Behaviour patterns, mechanisms and conservation implications. *Adv. Mar. Biol.* **2008**, *54*, 107–170. [PubMed]
55. Simpfendorfer, C.A. The Biology of Sharks of the Family Carcharhinidae from the Nearshore Waters of Cleveland Bay, with particular reference to *Rhizoprionodon taylori*. Doctoral Dissertation, James Cook University, Brisbane, Australia, 1993.
56. Hight, B.H.; Lowe, C.G. Elevated body temperatures of adult female leopard sharks, *Triakis semifasciata*, while aggregating in shallow nearshore embayments: Evidence for behavioral thermoregulation? *J. Exp. Mar. Biol. Ecol.* **2007**, *352*, 114–128. [CrossRef]
57. Jacoby, D.M.P.; Busawon, D.S.; Sims, D.W. Sex and social networking: The influence of male presence on social structure of female shark groups. *Behav. Ecol.* **2010**, *21*, 808–818. [CrossRef]
58. Elisio, M.; Colonello, J.H.; Cortés, F.; Jaureguizar, A.J.; Somoza, G.M.; Macchi, G.J. Aggregations and reproductive events of the narrownose smooth-hound shark (*Mustelus schmitti*) in relation to temperature and depth in coastal waters of the south-western Atlantic Ocean (38–42 S). *Mar. Freshw. Res.* **2016**, *68*, 732–742. [CrossRef]
59. Perry, C.T.; Clingham, E.; Webb, D.H.; de la Parra, R.; Pierce, S.J.; Beard, A.; Henry, L.; Taylor, B.; Andrews, K.; Hobbs, R.; et al. St. Helena: An important reproductive habitat for whale sharks (*Rhincodon typus*) in the central south Atlantic. *Front. Mar. Sci.* **2020**, *7*, 576343. [CrossRef]
60. Kock, A.; O'Riain, M.J.; Mauff, K.; Meÿer, M.; Kotze, D.; Griffiths, C. Residency, habitat use and sexual segregation of white sharks, *Carcharodon carcharias* in False Bay, South Africa. *PLoS ONE* **2013**, *8*, e55048. [CrossRef]
61. Becerril-García, E.E.; Hoyos-Padilla, E.M.; Micarelli, P.; Galván-Magaña, F.; Sperone, E. Behavioural responses of white sharks to specific baits during cage diving ecotourism. *Sci. Rep.* **2020**, *10*, 11152. [CrossRef]

62. Duffy, C.; Francis, M.P.; Manning, M.J.; Bonfil, R. Chapter 21: Regional population connectivity, oceanic habitat, and return migration revealed by satellite tagging of white sharks, Carcharodon carcharias, at New Zealand aggregation sites. In *Global Perspectives on the Biology and Life History of the White Shark;* Domeier, M.L., Ed.; CRC Press: Boca Raton, FL, USA, 2012; pp. 301–318.
63. Skomal, G.B.; Chisholm, J.; Correia, S.J. Implications of increasing pinniped populations on the diet and abundance of white sharks off the coast of Massachusetts. In *Global Perspectives on the Biology and Life History of the White Shark;* Domeier, M.L., Ed.; CRC Press: Boca Raton, FL, USA, 2012; pp. 405–418.
64. Robbins, R.L. Environmental variables affecting the sexual segregation of great white sharks *Carcharodon carcharias* at the Neptune Islands South Australia. *J. Fish Biol.* **2007**, *70*, 1350–1364. [CrossRef]
65. Semmens, J.; Payne, N.; Huveneers, C.; Sims, D.W.; Bruce, B. Feeding requirements of white sharks may be higher than originally thought. *Sci. Rep.* **2013**, *3*, 1471. [CrossRef]
66. Goldman, K.J.; Anderson, S.D. Space utilization and swimming depth of white sharks, *Carcharodon carcharias*, at the South Farallon Islands, central California. *Env. Biol. Fish.* **1999**, *56*, 351–364. [CrossRef]
67. Weng, K.C.; Boustany, A.M.; Pyle, P.; Anderson, S.D.; Brown, A.; Block, B.A. Migration and habitat of white sharks (*Carcharodon carcharias*) in the eastern Pacific Ocean. *Mar Biol* **2007**, *152*, 877–894. [CrossRef]
68. Dunbar, R.; Shultz, S. Evolution in the social brain. *Science* **2007**, *317*, 1344–1347. [CrossRef]
69. Clark, E. Instrumental conditioning of lemon sharks. *Science* **1959**, *130*, 217–218. [CrossRef]
70. Guttridge, T. The Social Organization and Behaviour of the Juvenile Lemon Shark *Negaprion breviostris*. Ph.D. Thesis, University of Leeds, Institute of Integrative and Comparative, Leeds, UK, 2009.
71. Northcutt, R.G. Elasmobranch central nervous system organization and its possible evolutionary significance. *Am. Zool.* **1977**, *17*, 411–429. [CrossRef]
72. Yopak, K.E.; Lisney, T.J.; Collin, S.P.; Montgomery, J.C. Variation in brain organization and cerebellar foliation in chondrichthyans: Sharks and holocephalans. *Brain Behav. Evol.* **2007**, *69*, 280–300. [CrossRef]
73. Striedter, G.F.; Northcutt, R.G. Head size constrains forebrain development and evolution in ray-finned fishes. *Evol. Dev.* **2006**, *8*, 215–222. [CrossRef] [PubMed]
74. Holland, K.N.; Meyer, C.G.; Dagorn, L.C. Inter-animal telemetry: Results from first deployment of acoustic 'business card' tags. *Endanger. Species Res.* **2009**, *10*, 287–293. [CrossRef]
75. Guttridge, T.L.; Gruber, S.H.; Krause, J.; Sims, D.W. Novel acoustic technology for studying free-ranging shark social behaviour by recording individuals' interactions. *PLoS ONE* **2010**, *5*, e9324. [CrossRef] [PubMed]
76. Krause, J.; James, R.; Croft, D.P. Personality in the context of social networks. *Philos. Trans. R. Soc. B Biol. Sci.* **2010**, *365*, 4099–4106. [CrossRef]
77. Klimley, A.P.; Nelson, D.R. Diel movement pattern of the scalloped hammerhead shark *Sphyrna lewini* in relation to el Bajo Espiritu Santo: A refuging central position social system. *Behav. Ecol. Sociobiol.* **1984**, *15*, 45–54. [CrossRef]
78. Sims, D.W.; Southall, E.J.; Quayle, V.A.; Fox, M. Annual social behaviour of basking sharks associated with coastal front areas. *Proc. R. Soc. Lond. Ser. B Biol. Sci.* **2000**, *267*, 1897–1904. [CrossRef] [PubMed]
79. Guttridge, T.L.; Gruber, S.H.; Gledhill, K.S.; Croft, D.P.; Sims, D.W.; Krause, J. Social preferences of juvenile lemon 543 sharks, Negaprion brevirostris. *Anim. Behav.* **2009**, *78*, 543–548. [CrossRef]
80. Jacoby, D.M.P.; Fear, L.N.; Sims, D.W.; Croft, D.P. Shark personalities? Repeatability of social network traits in a widely distributed predatory fish. *Behav. Ecol. Sociobiol.* **2014**, *68*, 1995–2003. [CrossRef]
81. Anderson, J.M.; Clevenstine, A.J.; Stirling, B.S.; Burns, E.S.; Meese, E.N.; White, C.F.; Logan, R.K.; O'Sullivan, J.; Rex, P.T.; May, J., III; et al. Non-random Co-occurrence of Juvenile White Sharks (*Carcharodon carcharias*) at Seasonal Aggregation Sites in Southern California. *Front. Mar. Sci.* **2021**, *8*, 688505. [CrossRef]
82. Papastamatiou, Y.P.; Mourier, J.; TinHan, T.; Luongo, S.; Hosoki, S.; Santana-Morales, O.; Hoyos-Padilla, M. Social dynamics and individual hunting tactics of white sharks revealed by biologging. *Biol. Lett.* **2022**, *18*, 20210599. [CrossRef]
83. Klimley, A.P.; Le Boeuf, B.J.; Cantara, K.M.; Richert, J.E. Automated tracking of white shark *Carcharodon carcharias* by radio-acoustic positioning system. *Mar. Biol.* **2000**, *4*, 182–193.
84. Hsu, Y.; Earley, R.L.; Wolf, L.L. *Modulating Aggression through Experience. Fish Cognition and Behaviour;* Blackwell: Oxford, UK, 2006; pp. 96–118.
85. Fallows, C.; Gallagher, A.J.; Hammerschlag, N. White sharks (*Carcharodon carcharias*) scavenging on whales and its potential role in further shaping the ecology of an apex predator. *PLoS ONE* **2013**, *8*, e60797. [CrossRef]
86. Bromilov, M. Feeding Behaviour of White Sharks Carcharodon carcharias around a Cage Diving Vessel and the Implications for Conservation. Ph.D. Thesis, University of Michigan, Department of Ecology & Evolutionary Biology, Ann Arbor, MI, USA, 2014; pp. 6–8.
87. Strong, W.R.; Murphy, R.C.; Bruce, D.B.; Nelson, D.R. Movements and associated observations of bait attracted white sharks *Carcharodon carcharias*: A preliminary report. *Mar. Freshw. Res.* **1992**, *43*, 13–20. [CrossRef]
88. Strong, W.R. Repetitive aerial gaping: A thwart-induced behaviour in white sharks. In *Great White Sharks;* Academic Press: Cambridge, MA, USA, 1996; pp. 207–215.

89. Lisney, T.J.; Bennet, M.B.; Collin, S.P. Volumetric analysis of sensory brain areas indicates ontogenetic shifts in the relative importance of sensory systems in elasmobranchs. *Raffles Bull. Zool.* **2007**, *55*, 7–15.
90. Reinero, F.R.; Sperone, E.; Giglio, G.; Pacifico, A.; Mahrer, M.; Micarelli, P. Influence of Environmental Factors on Prey Discrimination of Bait-Attracted White Sharks from Gansbaai, South Africa. *Animals* **2022**, *12*, 3276. [CrossRef] [PubMed]

diversity

MDPI

Article

A Cocktail of Plankton and Organochlorines for Whale Shark in the Foraging Areas of Nosy Be (Madagascar)

Letizia Marsili [1], Guia Consales [1,*], Patrizia Romano [1], Rachele Rosai [1], Paolo Bava [2], Francesca Romana Reinero [2] and Primo Micarelli [1,2]

1 Department of Physical Sciences, Earth, and Environment, University of Siena, Via Mattioli, 4, 53100 Siena, Italy
2 Sharks Studies Center—Scientific Institute, 58024 Massa Marittima, Italy
* Correspondence: guia.consales@unisi.it

Abstract: Seas and oceans are contaminated by persistent organic pollutants (POPs), which are released into the environment by human activities. The chemical-physical properties of POPs induce high persistence and toxicity in marine organisms from the lowest to the highest trophic levels. Phyto- and zooplankton are at the base of the food chain, and they can adsorb and accumulate these xenobiotic compounds. Therefore, all planktophagous species, including the whale shark (*Rhincodon typus*), are susceptible to ingesting these contaminants during feeding. From October to December, whale sharks migrate along the north-west coast of Madagascar in search of dense patches of plankton. During scientific expeditions to the whale sharks' foraging areas in the waters of the island of Nosy Be (which is in the north-west of Madagascar), plankton samples were taken. In these samples, the presence and levels of some chlorinated xenobiotics (HCB, DDT and its metabolites, and PCBs) were evaluated in order to estimate the possible impact of whale shark diet on organochlorine (OC) accumulation. The fresh plankton biomass sampled from this region did not seem to be sufficient for the sustenance of the animals, which suggests that the daily contamination input of *Rhincodon typus* individuals, depending on their plankton diet, is minimal.

Keywords: zooplankton; pollution; legacy contaminants; POPs; DDTs; PCBs; HCB; *Rhincodon typus*; contaminant intake

1. Introduction

Zooplankton plays an important role in regulating the patterns and mechanisms through which both matter and energy are transferred from the base to the upper levels of food webs [1]. Zooplankton is a key vehicle through which persistent contaminants entering the marine environment are transferred from primary producers to higher trophic levels, since it accumulates pollutants from both water and food [2]. Zooplankton provides an essential food source for numerous species, and its fluctuations in spatio-temporal distribution might influence the biodiversity trends in various marine organisms [3], including whale sharks.

The whale shark, *Rhincodon typus* (Smith, 1828), is the largest known fish-like vertebrate in the world, with an uncertain maximum size [4]. It has been included in the CITES Appendix II since 2002. The species was also listed as vulnerable in the IUCN Red List in 2000 [5], a status that was confirmed in 2005 [6], and in 2016, the conservation status was then reclassified as endangered in order to address its decreasing population trend [7]. Whale sharks are panoceanic planktivores and are cosmopolitan in distribution, inhabiting all tropical and warm temperate seas, except the Mediterranean [8]. They spend, on average, 7.5 h/day feeding at the surface on dense plankton dominated by calanoids, copepods, sergestids, chaetognaths, and fish larvae [9–13]. The filtering apparatus of *R. typus*, unlike that of *Cetorhinus maximus* and *Megachasma pelagios*, is incapable of filtering large volumes of water, but it seems to be adapted to a combination of filter and suction feeding, making it more versatile than that of the other two filtering sharks, and thereby allowing whale sharks to target a wider variety of prey [14].

Most of the information on the species of the whale shark is derived from studies conducted in coastal areas [15,16], where, seasonally, various individuals aggregate based on environmental factors, such as the seasonal productivity of plankton [17–20], the reproduction in fish [21], crab egg releases [22,23], and ocean current trends [24]. The discovery of several aggregation sites of these animals around the world, including in the waters of Nosy Be Island in Madagascar, has significantly increased the number of sightings in recent years, and in many areas, these encounters have helped to develop a profitable and increasingly popular tourism industry [8,25–28].

However, increasing human activity in whale shark feeding grounds has, in turn, increased chemical pollution from urban wastewaters, vessels, agriculture, and also waste. Primary information regarding contaminant uptake in elasmobranch species is still lacking, though, as well as the potential physiological effects of pollutants on the whale shark species [20,29]. As a result, even if pollution has not yet been considered as one of the main threats to the survival of the whale shark species, the negative effects on the health of this organism may worsen the situation [30,31]. Contaminants entering the marine environment are readily absorbed by organic matter, and they are taken up and absorbed by plankton at the base of marine food webs [32,33]. Marine zooplankton has relatively high lipid reserves, and it can accumulate hydrophobic organochlorine compounds (OCs) [34–36]. Several studies have shown that sharks bioaccumulate and biomagnify certain metals and metalloids in their tissues as well as organochlorine contaminants [2]. Although POPs have been regulated since the 1970s and banned in production and in use by the 2001 Stockholm Convention (initially there were 12, called the "dirty dozen", but there are now a total of 29, plus another 6 that are under review) [37], large quantities of persistent organic pollutants (POPs) have been released into the environment. Due to their propensity for long-range transport, high environmental persistence, bioaccumulation potential, and intrinsic toxicity, POPs continue to present a global problem today [38]. Dichlorodiphenyltrichloroethane (DDT) and polychlorinated biphenyls (PCBs), for instance, are still found in countries within the Northern Hemisphere, where they have actually been banned already for a long period of time [2]. Contamination of marine environments has been linked to increasing levels of lethal and sub-lethal effects to individuals, populations, and ecosystems [24,39]. Chronic or intermittent exposure to OCs and trace elements results in severe effects on aquatic organisms at different physiological, cellular, and behavioral levels [24]. In this study, zooplankton samples were collected between November and December of 2019 along the coast of Nosy Be Island in the foraging areas of whale sharks. The samples were analyzed in order to evaluate the presence and the levels of OCs, particularly hexachlorobenzene (HCB), 29 PCB congeners, and DDT with its metabolites (DDTs). Knowing the feeding habits of the whale shark [9] and the amount of fresh plankton biomass present in this area at the time of the elasmobranch visitation [12], it was also possible to evaluate the input of organochlorine contamination through the animal's diet.

2. Materials and Methods

2.1. Study Area

Nosy Be (≈13°39′ S; 40°20′ E), in the Antsiranana Province, is a volcanic island located in the Mozambique Channel, 8 km (km) off the Northwest coast of Madagascar (Figure 1). The island is roughly 22.5 km long and 15 km wide with an area of 312 square km, and Mont Lokobe is its highest peak at 450 m. Water depths on the continental platform around Nosy Be are generally shallow, rarely exceeding 40 m. Water temperatures around Nosy Be vary from about 24 °C in August to about 28 °C in February. The difference between the lowest and highest possible tides is 4.44 m, with an average of 2.22 m. This great fluctuation in tidal level gives rise to strong tidal currents in restricted channel areas [40]. Nosy Be is a small island of Madagascar, famous for its own largely endemic animal and plant biodiversity [41]. The distribution, status, and abundance of whale sharks are poorly documented in Madagascar [42,43]. North-western Madagascar is a significant hotspot for marine megafauna species, including whale sharks, cetaceans, and sea turtles, as well as for coral biodiversity [44]. Nosy Be, specifically, is likely to be a feeding area for planktivores: *R. typus* here is often associated with surface schools of mackerel tuna feeding on small pelagic fishes (*Clupeidae*) [45]. The population structure of whale sharks, the majority of which are juvenile males, is common within their coastal feeding areas [46]. Among the various impacts in Nosy Be and the larger Madagascar area, both anthropogenic (fishing methods, uncontrolled tourism, and recreational activities) and environmental (tropical cyclones—from November to April—and climate change effects generally) have amplified over the last four decades, and the stress related to increased pollution associated with dredging, coastal development, deforestation, and intensive agriculture should not be overlooked [45]. This is particularly the case for those species, such as the whale shark, which are already considered at risk, and that will be in need of conservation in the near future.

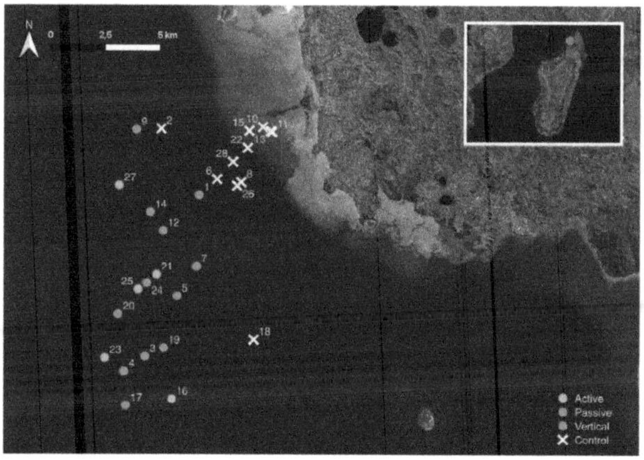

Figure 1. Study area (south-eastern Nosy Be, in north-western Madagascar). Green dots represent plankton sampled during active feeding behavior of whale sharks. Orange dots represent plankton sampled during passive feeding behavior, and pink dots represent plankton sampled during vertical feeding behavior. White "X"s represent plankton sampled when no whale sharks were present. Numbers near the dots represent sample IDs.

2.2. Samples Collection

One expedition, coordinated by the Sharks Studies Centre, was carried out in November–December of 2019 with the logistical support of Manta Diving boats. Departing from the beach of Ampasikely (S 13°20′47.9335″ E 48°11′19.824″) on small tourist boats, the team collected zooplankton samples in the coastal waters of the south-eastern side of the island from approximately 08:30 to 13:30. The plankton samples were collected with a 200 µm mesh size net with a 50 cm diameter mouth. Once the net had been ballasted (2–5 kg) and tied with a rope to the boat, it was towed for 10 min at a speed of 2 knots. Afterwards, the net was recovered and repeatedly rinsed with sea water from the outside to the inside in order to recover any material stuck to its walls. Plankton entrapped in the collector at the base of the net was then retrieved and filtered through a 0.50 µm filter, wrapped in a sheet of aluminum foil using a spatula, placed inside plastic jars, and, finally, stored in a refrigerator. Zooplankton collected in the feeding area of the whale sharks was sampled during or immediately after the sharks displayed a feeding behavior (active, passive, and vertical), following the definition by [47]. Furthermore, upon returning from each sampling trip, a control sample was taken near the coast where, according to the local guides and fishermen, whale sharks are never encountered. This was conducted at a depth of at least 10 m in order to avoid breaking the net along the seafloor. Upon returning from the survey, each sample was divided into sub-samples for the different analyses. A part of the sample was used for taxonomic investigations [12], and another aliquot was frozen at −20 ° C for toxicological analyses. A total of 28 zooplankton samples were analyzed for OC determination (Figure 1).

2.3. Sample Preparation and OC Determination

Frozen plankton samples were weighed. All samples were <1 g, so a standard weight of 0.100 g was used for OC determination. HCB, DDTs, and PCBs were determined according to U.S. Environmental Protection Agency (EPA) method 8081/8082 modified according to Marsili et al. (2016) [39]. Samples were then freeze-dried for 2 days, and were subsequently homogenized manually with a mortar. Cellulose thimbles were pre-extracted in a Soxhlet for 9 h with 150 mL of n-hexane in order to remove impurities. Next, the thimbles were evaporated under a fume hood for one hour and placed in the stove at 100 °C for another hour. A total of 0.100 g of the plankton samples were loaded into each cellulose thimble, spiked with 100 µL 1 ng/µL of PCB n° IUPAC30 (International Union of Pure and Applied Chemistry) [47], and then extracted in a Soxhlet apparatus for 9 h with 200 mL of n-hexane. The extracted organic material (EOM%; lipid content) was calculated gravimetrically in each sample.

The extract was then saponified with 10 mL of sulphuric acid (98% AnalR © Normapur, VWR chemicals) for 12 h to obtain lipid sedimentation. Supernatant solution was recovered and evaporated to 10 mL with Rotavapor 110 at constant temperature of 45 °C. The extract then underwent liquid chromatography on a column containing Florisil (VWR chemicals, ph 8.5; mesh size 150–250 µm) that had been dried at 110 °C for 1 h, and everything was eluted with 90 mL of n-hexane. This phase further purified the apolar phase of the lipids that could not be saponified. The extract was then evaporated and spiked with 100 µL of hexane 0.15 ng/µL of PCB209, which was used as the second internal standard.

The analytical method used was High Resolution Capillary Gas Chromatography with an Agilent 6890 N and a 63 Ni ECD and an SBP-5 bonded phase capillary column (30 m long, 0.2 mm internal diameter). The carrier gas was nitrogen with a head pressure of 15.5 psi (splitting ratio 50/1). The scavenger gas was argon/methane (95/5) at 40 mL/min. The oven temperature was 100 °C for the first 10 min, after which it was increased to 280 °C at 5 °C/min. The injector and detector temperatures were 200 and 280 °C, respectively. A mixture of specific isomers was used to calibrate the system, evaluate recovery, and confirm the results.

The standard injected was prepared with 50 ng/mL of HCB, 100 ng/mL of DDT (pp′DDT, pp′DDD, pp′DDE, op′DDD, op′DDE), 200 ng/mL of op′DDT, and 2 µg/mL of Arochlor 1260. For the evaluation of the linearity in the instrumental response and the instrumental sensitivity, the following quantities of the standard were injected: 1, 2, and 4 µL. Capillary gas chromatography revealed 29 PCB congeners (IUPAC no. 95, 99, 101, and 118—pentachlorobiphenyls; 128, 135, 138, 144, 146, 149, 151, 153, and 156—hexachlorobiphenyls; 170, 171, 172, 174, 177, 178, 180, 183, and 187—heptachlorobiphenyls; 194, 195, 196, 199, 201, and 202—octachlorobiphenyls; and 206—nonachlorobiphenyls). Total PCBs (∑PCBs) were quantified as the sum of all congeners. Total DDTs (∑DDTs) were calculated as the sum of the isomers op′DDT, pp′DDT, op′DDD, pp′DDD, op′DDE, and pp′DDE. The limit of detection (LOD) for all compounds analyzed was 0.1 ng/kg (ppt).

2.4. Data Analysis

Data were processed with STATISTICA 7.1 software. A Shapiro–Wilk test was used to check the distribution of the data. The Shapiro–Wilk test utilizes the null hypothesis principle: the null hypothesis is that the population is normally distributed ($p > 0.05$). In the non–normally distributed data, a Kruskal–Wallis test was applied, and in those normally distributed or normalized with Log transformation, a t-test and a Pearson test were applied.

3. Results and Discussion

3.1. POP Concentrations in Plankton Samples

Table 1 summarizes POP concentrations, expressed in ng/g dry weight (d.w.), detected in plankton samples, and divided between the functions of the shark feeding behavior (active (A), passive (P), vertical (V)), and the control area (C)) where the shark was not present.

Table 1. HCB, PCB, and DDT levels in plankton samples divided by the feeding behavior of the whale shark (vertical, passive, and active). Control refers to plankton sampled when no shark was around. N = number of samples; SD = Standard Deviation; Min = Minimum; Max = Maximum; SE = Standard Error. All values are expressed in ng/g dry weight.

		Vertical		
Compound	**N**	**Mean ± SD (Min–Max)**	**Median**	**SE**
HCB	7	2.80 ± 4.02 (0.47–10.9)	1.35	1.64
PCBs	7	114.55 ± 160.04 (26.16–436.49)	47.39	65.33
DDTs	7	44.20 ± 29.26 (15.17–93.57)	34.65	11.95
DDTs/PCBs	7	0.64 ± 0.22 (0.21–0.83)	0.64	0.09
		Passive		
Compound	**N**	**Mean ± SD (Min–Max)**	**Median**	**SE**
HCB	5	1.65 ± 1.05 (0.70–2.93)	1.08	0.47
PCBs	5	65.81 ± 25.68 (38.95–98.05)	57.16	11.48
DDTs	5	44.40 ± 21.06 (23.65–74.98)	37.27	9.42
DDTs/PCBs	5	0.65 ± 0.06 (0.61–0.76)	0.64	0.03
		Active		
Compound	**N**	**Mean ± SD (Min–Max)**	**Median**	**SE**
HCB	5	0.72 ± 0.19 (0.51–1.01)	0.68	0.09
PCBs	5	41.07 ± 16.10 (28.55–67.91)	34.69	7.20
DDTs	5	25.85 ± 6.02 (18.11–32.96)	26.28	2.70
DDTs/PCBs	5	0.68 ± 0.22 (0.44–0.95)	0.63	0.10
		Control		
Compound	**N**	**Mean ± SD (Min–Max)**	**Median**	**SE**
HCB	11	0.89 ± 0.81 (0.41–3.31)	0.68	0.24
PCBs	11	47.76 ± 14.49 (27.99–73.62)	49.74	4.37
DDTs	11	30.21 ± 10.16 (14.54–54.46)	26.81	3.06
DDTs/PCBs	11	0.65 ± 0.20 (0.40–1.09)	0.60	0.06

In the (V) samples, the mean levels of HCB and PCBs were higher than in the other samples; DDTs had levels comparable to those of sample (P), and both were higher than (A) and (C). The differences between the four groups, however, were not statistically significant ($p < 0.05$) with the non-parametric Kruskal–Wallis H test. This is probably due to the low sample number evaluated and the high standard deviation that was detected. The average abundance pattern for the target contaminants in plankton was PCBs > DDTs > HCB, regardless of the type of feeding behavior and the sampling area. Considering all of the zooplankton samples together, PCBs ranged from 26.16 ng/g d.w. to 436.49 ng/g d.w. ($\bar{x} = 64.71 \pm 77.14$), DDTs ranged from 14.54 ng/g d.w. to 93.57 ng/g d.w. ($\bar{x} = 35.14 \pm 18.46$), and HCB ranged from 0.41 ng/g d.w. to 10.90 ng/g d.w ($\bar{x} = 1.42 \pm 2.05$). Evaluating the representativeness of these results from a quantitative point is challenging, especially because sources for bibliographic comparison are scarce. There are very few previous studies on OC levels in plankton globally (Table 2), and only one was conducted near the wide-ranging area covered by this study [2].

Table 2. Bibliographic research on studies in which PCBs, DDTs, and HCB were evaluated in phyto- and zooplankton all over the world. Values are expressed in mean ± SD or as a range of minimum-maximum. w.w. = wet weight; d.w. = dry weight; l.w. = lipid weight.

Area	Sample Type	PCBs	DDTs	HCB	Ref.
Gulf of Mexico and Caribbean	Zooplankton	<3–678 ng/g w.w.	0.2–34 ng/g w.w.		[48]
Turku Arcipelago (Finland)	Zooplankton	38 ppm l.w.			[49]
Southern Ocean	Zooplankton and phytoplankton	0.30–0.37 ng/g d.w.	19 ng/g d.w.		[50]
Terranova Bay (Antartide)	Zooplankton (copepods)	575 ng/g l.w.	400 ng/g l.w.	109 ng/g l.w.	[51]
East coast of Newfoundland (Canada)	Zooplankton	85.7 ng/g l.w.	22.3 ng/g lw	6.4 ng/g l.w.	[52]
Pelagos Sanctuary (Maditerranean Sea)	Zooplankton *(Meganyctiphanes norvegica)*	84.6–210.2 ng/g w.w.	45.3–163.2 ng/g w.w.	3.5–11.6 ng/g w.w.	[53]
Portugal	Plankton	61–159 ng/g d.w. (February) 68–155 ng/g d.w. (April) 12–63 ng/g d.w. (July)	48–76 ng/g d.w. (north) 3–7 ng/g d.w. (south)		[54]
Maditerranean Sea	Zooplankton	0.76–353 ng/g d.w.		2.5 ng/g d.w.	[55]
Strait of Georgia British Columbia (Canada)	Zooplankton	52.2–364 ng/g l.w.			[56]
Coastal Transect in British Columbia (Canada)	Zooplankton	0.2–0.8 ng/g l.w. (north) 0.6–1.2 ng/g l.w. (south)			[57]
Atlantic, Indian and Pacific Oceans	Zooplankton	30–692 pg/g d.w.			[58]
Gulf of Tadjoura (Djibouti)	Zooplankton	109.7–636.1 ng/g d.w.	21.42–79.2 ng/g d.w.		[2]
Weizhou Island (China)	Zooplankton		0.77 ± 0.20 ng/g d.w.	0.20 ± 0.08 ng/g d.w.	[59]

The results obtained in our study were in line with those conducted in Portugal [54], while those recorded in the Southern Ocean [50] were considerably lower. The most interesting comparison is with the study conducted in Djibouti [2], which is 3000 km north of Nosy Be, in which PCB levels were higher despite the similarity of the area. On the other hand, DDTs were consistent with the aforementioned studies [2,50,54].

The only two studies in which HCB analysis was carried out were those conducted in the Mediterranean Sea [55] and in Weizhou Island [59], and in both cases, our results were higher.

3.2. PCB Congeners Composition

The PCB content was mostly dominated by eight congeners: PCB(149 + 118), PCB153, PCB138, PCB180, PCB170, PCB201, and PCB206, with contributions > 50% (Figure 2).

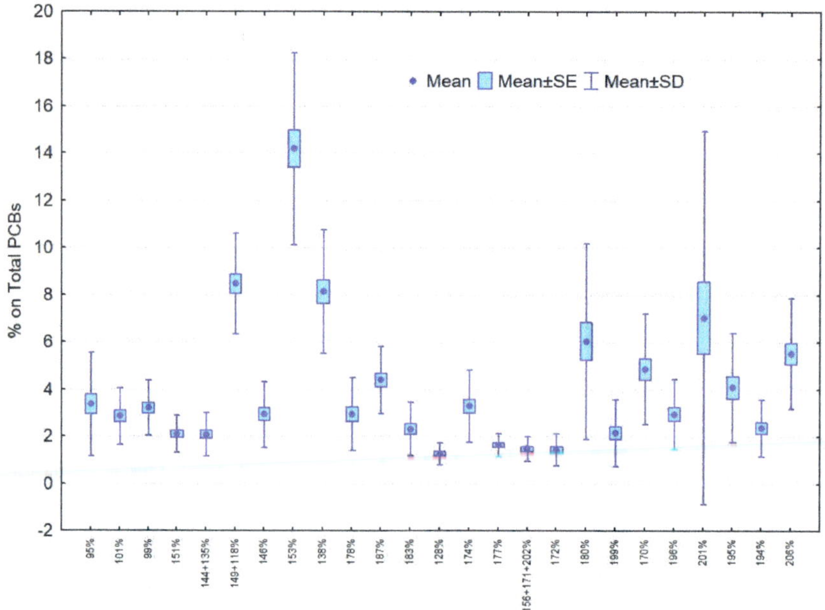

Figure 2. PCB fingerprint registered in all of the plankton samples. Values are expressed in percentage (%) on total PCBs. X-axis represents the 29 PCB congeners detected in the plankton samples.

It should be emphasized that the PCB abundance model is very similar to what is usually reported for biotic matrices, where the most recalcitrant PCB congeners (in particular, PCB153, PCB138, and PCB180) constituted the majority of PCB burdens. Among these, congener 22′44′55′ (PCB153) was the most abundant in all of the samples, probably due to the fact that this congener is particularly persistent as it has chlorines in positions 2, 4, and 5 of both rings of the biphenyl [60–62] and has no adjacent unsubstituted carbons in the ortho-meta position [63]. It is also very important in toxicological terms as a mutagenic, teratogenic, and carcinogenic compound, and also as an endocrine disruptor [64]. PCB118—which was present in high percentages in all of the samples—is also important toxicologically because it shares the previously mentioned characteristics [64]. Figure 2 shows the very high SDs, particularly for the congeners PCB153, PCB180, and PCB201. Therefore, we investigated whether the differences in feeding behavior type could be related to a different PCB fingerprint between the groups. With the *t*-test, performed following the verification of the Gaussian distribution of the data and considering the percentage values, significant differences ($p < 0.05$) were identified: (V) differs from (P) for PCB178 and PCB196 as well as from (A) for PCB172, and only in the case of PCB196 we had the highest percentages in (V); (P) differs from (A) and from (C) for PCB99, always in lower percentages in (P); (P) differs from (A) for PCB151 and from (C) for PCB178, and PCB178 were present with higher percentages in (P), unlike PCB151; and, finally, (A) had significantly higher percentages of PCB172 and PCB199 than (C). Thus, only a few congeners differ significantly between the different feeding behaviors, and among these, the most representative PCB congeners were not present. This result could have been influenced by the high SD recorded within each group for the individual samples that were analyzed. For this purpose, the PCB fingerprint results are reported separately for the four behavioral groups (Table 3). Analyzing the data, PCB180 has a high SD both in (V) and in (P), PCB201 has a high SD in both (V) and in (C), and even in (C), the SD for PCB201 is greater than the mean value.

Table 3. PCB fingerprint registered in the plankton samples divided by feeding behavior. Values are expressed in mean percentage (%) on total PCBs ± Standard Deviation.

Compound	Vertical N = 7	Passive N = 5	Active N = 5	Control N = 11
95	3.48 ± 1.46	2.74 ± 2.41	3.26 ± 0.59	3.61 ± 2.94
101	3.14 ± 1.92	2.68 ± 1.49	2.81 ± 1.04	2.78 ± 0.71
99	3.44 ± 1.60	2.02 ± 0.75	3.73 ± 0.94	3.37 ± 0.92
151	2.28 ± 1.16	1.61 ± 0.55	2.45 ± 0.44	2.02 ± 0.70
144 + 135	1.89 ± 1.10	1.69 ± 0.67	2.32 ± 0.43	2.25 ± 1.08
149 + 118	8.40 ± 2.62	8.01 ± 3.18	8.40 ± 2.10	8.75 ± 1.53
146	2.57 ± 0.91	2.63 ± 2.77	3.02 ± 1.37	3.23 ± 0.79
153	14.82 ± 3.38	15.16 ± 3.56	13.80 ± 1.25	13.59 ± 5.51
138	9.18 ± 3.01	9.14 ± 1.81	7.36 ± 1.22	7.45 ± 3.07
178	2.38 ± 0.87	4.09 ± 1.16	3.33 ± 2.45	2.54 ± 1.34
187	4.43 ± 1.02	5.47 ± 2.04	4.32 ± 0.66	3.94 ± 1.49
183	2.05 ± 0.43	2.85 ± 2.34	1.98 ± 0.58	2.38 ± 0.84
128	1.37 ± 0.48	1.61 ± 0.78	1.16 ± 0.26	1.10 ± 0.29
174	2.85 ± 1.01	3.90 ± 1.46	4.05 ± 2.52	2.90 ± 1.22
177	1.75 ± 0.49	1.76 ± 0.55	1.71 ± 0.51	1.46 ± 0.46
156 + 171 + 202	1.55 ± 0.39	1.93 ± 0.84	1.24 ± 0.42	1.32 ± 0.30
172	1.23 ± 0.40	1.41 ± 0.74	2.34 ± 0.85	1.23 ± 0.44
180	7.66 ± 6.94	8.02 ± 5.26	5.06 ± 1.12	4.70 ± 1.75
199	2.17 ± 1.24	2.49 ± 2.67	3.15 ± 1.06	1.53 ± 0.40
170	5.54 ± 2.94	5.51 ± 3.28	3.77 ± 2.09	4.68 ± 1.69
196	3.40 ± 0.70	2.09 ± 0.52	2.62 ± 0.97	3.23 ± 2.10
201	4.43 ± 3.34	5.20 ± 2.07	4.31 ± 1.54	10.56 ± 11.44
195	2.85 ± 1.82	4.02 ± 3.43	4.61 ± 2.26	4.44 ± 2.39
194	2.29 ± 0.96	1.95 ± 1.19	2.97 ± 1.17	2.31 ± 1.40
206	5.31 ± 2.91	4.53 ± 2.68	6.69 ± 1.51	5.48 ± 2.31

3.3. DDT Isomer Composition and Ratios

The relative contribution to the total DDT content (Figure 3) was pp′DDE (43.2%) > op′DDT (23.8%) > pp′DDT (13.2%) > op′DDD (7.3%) > pp′DDD (6.7%) >op′DDE (6.4%). For the first three isomers (pp′DDE, op′DDT, and pp′DDT), a similar trend was observed in the behavioral groups, while minimal differences existed for op′DDD, pp′DDD, and op′DDE (Table 4). The only significant difference between the groups was between (V) and (A) for pp′DDT, with a higher percentage in group (A). Typically, technical DDT is composed of pp′DDT (77.1%), op′DDT (14.9%), pp′DDD (0.3%), op′DDD (0.1%), pp′DDE (4.0%), op′DDE (0.1%), unidentified compounds (3.5%) [65], and a (pp′DDE/pp′DDT) ratio of 0.05. If the ratio (pp′DDE/pp′DDT) has high values, it can be deduced that the majority of the active substance (pp′DDT) has been degraded to pp′DDE, and, therefore, there were no recent deposits of insecticide into that ecosystem [66]. In all the zooplankton samples the pp′DDE/pp′DDT ratio had a mean value of 4.55, with a range from 0.78 to 12.97. Clearly, this is a value higher than 0.05 of the technical DDT, but is not so high as to suggest an historic introduction of the pesticide. The (pp′DDE/DDTs) ratio, as well as having a similar meaning to the (pp′DDE/pp′DDT) ratio, can also indicate the efficiency of the metabolic processes [67]. In fact, the (ppDDE/DDTs) ratio indicates the relative abundance of metabolized forms of DDT. In this study, zooplankton showed a ratio that varied from 0.23 to 0.65, with a mean of 0.43. A value of this ratio equal to 0.6 is considered critical, while higher values indicate that there are no new contamination inputs in the study area [68]. The value found in our samples thus highlighted an alarming situation.

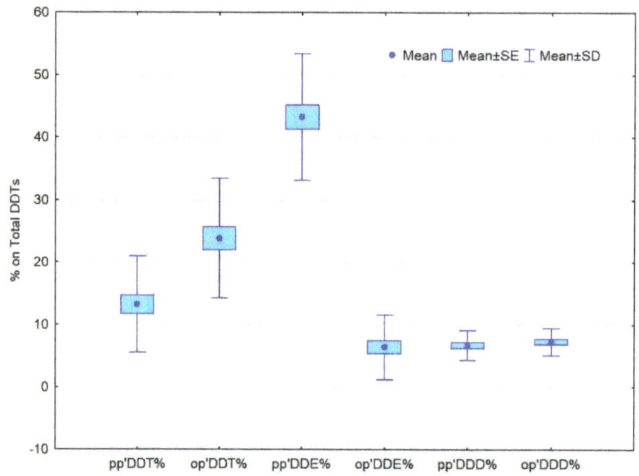

Figure 3. DDT fingerprint registered in all the plankton samples. Values are expressed in percentage (%) on total DDTs.

Table 4. DDT fingerprint (expressed in percentage (%) on total DDTs) and DDT isomer ratios registered in the plankton samples divided by feeding behaviour. op′DDTs = (op′DDD + op′DDE + op′DDT).

Compound	Vertical N = 7	Passive N = 5	Active N = 5	Control N = 11
op′DDE	6.11±4.59	8.84 ± 7.45	2.98 ± 1.17	7.25 ± 5.15
op′DDD	8.10 ± 2.43	5.41 ± 2.30	7.94 ± 1.89	7.45 ± 1.79
op′DDT	20.86 ± 10.12	24.91 ± 13.42	20.01 ± 5.30	26.66 ± 9.15
pp′DDE	49.45 ± 9.90	39.78 ± 9.25	42.49 ± 8.21	41.73 ± 11.25
pp′DDT	8.59 ± 3.60	13.95 ± 8.96	20.20 ± 11.85	12.27 ± 4.57
pp′DDD	6.89 ± 1.61	7.12 ± 3.00	6.38 ± 2.36	6.62 ± 2.80
pp′DDE/pp′DDT	6.91 ± 3.68	4.39 ± 3.67	2.89 ± 1.91	4.08 ± 2.43
(pp′DDE + pp′DDD)/pp′DDT	7.80 ± 3.96	5.26 ± 4.52	3.30 ± 2.18	4.68 ± 2.59
pp′DDE/DDTs	0.49 ± 0.10	0.40 ± 0.09	0.42 ± 0.08	0.42 ± 0.11
op′DDTs/DDTs	0.35 ± 0.09	0.39 ± 0.11	1.47 ± 1.33	0.40 ± 0.10
op′DDT/pp′DDT	2.91 ± 2.45	2.70 ± 2.10	1.47 ± 1.33	2.45 ± 1.20

Another ratio used as an indicator of fresh or altered residues [69] was that between the sum of pp′DDE and pp′DDD on the pp′DDT [(pp′DDE + pp′DDD)/pp′DDT].

Normally, a value of 1 is taken to distinguish between legacy and recent DDT inputs [70]. The mean value found for all of the samples was 5.23, with a range in different feeding behavior from 3.30 (A) to 7.80 (V). A total of 100% of the 28 samples had a value of this ratio > 1. These results suggest the recent use of DDT in the Nosy Be area, likely illegally. This theory is further supported by the relationship that is seen between the op′ isomers of DDT and DDTs [(op′DDE + op′DDD + op′DDT)/DDTs]. A sum of op′ isomers that exceed 20% of total DDT suggests a non-insecticide (or industrial) source of this xenobiotic [71]. In fact, the waste products from the processing of technical DDT are generally enriched with op′ isomers with respect to pp′DDT: the resulting compound finds application on an industrial level, and it is not subject to regulation for the use of DDT insecticide mixtures [72]. The value for all of the samples was 0.37, with a range of 0.21–0.56. The same results were obtained by separating the four types of sampling with the values of this ratio that were higher than 20% (Table 4), and this suggests that there is an excess of op′ isomers used in DDT mixtures, which possibly indicates that the xenobiotic is of industrial origin. Furthermore, the (op′DDT/pp′DDT) ratio is considered a discriminating indicator between the use of Technical DDT and dicofol [69,73]. The latter is a miticidal pesticide and acaricide synthesized from DDT and, thus, the isomers op′DDT, op′DDE, pp′DDT, or pp′-Cl-DDT (1,2,2,2-tetrachloro-1,1-bis(4-chlorophenyl)ethane—a chlorinated DDT intermediate that leads to dicofol prior hydrolysis)—are usually found in formulations of this pesticide [60,74]. Dicofol is very toxic to aquatic organisms, and it is highly bioaccumulative and degrades moderately slowly in both soil and sediments. It can be identified as a POP in terms of its long-range transport potential exhibiting a higher Arctic contamination potential [75]. Dicofol is also known to be neurotoxic and to possess endocrine disrupting properties, and this is the case both as an original product and with decomposition products [76]. Dicofol was listed in Annex A of the Stockholm Convention on Persistent Organic Pollutants only after the ninth meeting of the Conference of the Parties held in 2019 [77].

The concentration range of DDT impurities can vary widely. The total DDT content was found between 0.3% and 14.3% of the total weight of dicofol [78], although dicofol produced in China was reported to have on average 20% of DDT [73,79]. Given that in technical DDT, it is typically ~0.19 [80], and given that op′DDT shows a shorter half-life than pp′DDT in the environment [81], it seems reasonable to assume the influence of dicofol-type contamination when encountering an (op′DDT/pp′DDT) value > 0.2. In these samples, the mean value of this ratio was 2.42, ranging from 0.36 to 7.79. Values of this ratio > 0.2 were observed in 37% of the samples, and this indicates possible dicofol contamination. After applying the data log transformation to meet the criterion for normality, an attempt was made to evaluate whether there was a correlation between the [(pp′DDE+pp′DDD)/pp′DDT] and (pp′DDE/pp′DDT) ratios, which would have further confirmed the possible presence of dicofol. This is always linked to the degradation of the pp′-Cl-DDT impurity of dicofol, which, by degrading into pp′DDE, would increase the values of the ratios. The significant Pearson correlation coefficient (Figure 4) suggests this as a possible route of contamination for the analyzed zooplankton.

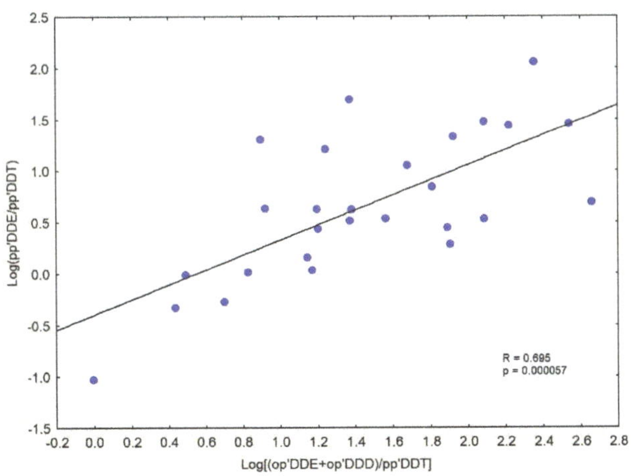

Figure 4. Pearson's correlation between [(op′DDE+op′DDD)/pp′DDT] and (pp′DDE/pp′DDT) in all the analysed plankton samples. $R = 0.695$; $p < 0.0001$.

Finally, we evaluated the relationship between DDTs and PCBs, the ratio of which (DDTs/PCBs) is indicative of contamination that is more likely to be of agricultural origin if the value is >1, and more likely to be of industrial origin if the value is <1 [82]. A total mean value for 28 samples of 0.65 with a range in the different groups between 0.64 (V) and 0.68 (A) indicates that in the waters of Nosy Be, where the plankton were sampled, the greatest impact is likely to be generated from the industrialized areas. This seems to contrast with the fact that agriculture (cotton, tobacco, coffee, cacao, and cinnamon, as well as other spices) is the main activity in the area and makes the largest contribution to Madagascar's economy, employing about 85% of the population, followed by tourism, the production of goods with low added value, and then the mining sector [83]. However, the area surrounding Madagascar, which is very important for the richness of biodiversity, is considered as one of the areas with the greatest human impact [84].

3.4. Potential POPs Uptake in the Whale Shark

Three samples collected in the control area and three in the feeding area for each of the feeding behaviors were analyzed for mesozooplankton composition, and the results were reported in Bava et al. [12]. No significant differences were found between the feeding and the control areas. The total number of individuals was 472.44 $\pm$ 44.32 ind/m^3 (mean $\pm$ S.E.). The most common taxonomic group was Copepoda, followed by Appendicularia, Mollusca, and Chaetognatha. Biomass was calculated for each sample with the method of Di Capua et al. [85]. Wet biomass was 30.51 $\pm$ 3.57 mg/m^3, and 80% was represented by size class $\leq$ 2 mm, mostly by Copepoda. Wet biomass was higher in the control area compared to the feeding area, probably due to the absence of predator pressure, including the whale shark. However, this difference was not statistically significant. In the Nosy Be area at the time of zooplankton sampling, 48 whale sharks of approximately 4 m (m) were identified. On average, a whale shark about 4 m in length filters 326 m^3/h of water [9]. In a habitat rich in planktonic species, such as Cabo Catoche (Mexico), which has a fresh plankton biomass of 4.5 g/m^3, a 4 m long individual collects about 1467 g/h of plankton. By filtering for 7.5 h/day on the surface, it takes in about 11,002.5 g of plankton per day, which at 1.357 kJ/g corresponds to 14,931 kJ/day (3569 kcal/day) [9]. Considering that the total fresh biomass found in Nosy Be is about 30.51 mg/m^3, a 4 m whale shark could ingest about 74.60 g of plankton per day, which would correspond to about 101.23 kJ/day (24.19 kcal/day). This calculation allows us to deduce that in the waters of the island of Nosy Be, the individuals of *R. typus* cannot obtain the energy necessary to maintain their body biomass from plankton alone, and this is insufficient for their survival. In the feeding grounds in which whale sharks were identified in Nosy Be, in addition to plankton, there are also tuna, anchovies, mackerels, and other small nektonic species, and, as suggested by Diamant et al. [86], based on routine visual observations of sharks following—and occasionally successfully feeding on—bait fish, they could be the primary target when the sharks are near Nosy Be. There are not many studies in the literature that demonstrate this hypothesis, with the exception of the work by Boldrocchi and Bettinetti [87] carried out in the Gulf of Tadjoura (Djibouti), where whale sharks were filmed while feeding on a school of anchovies, probably belonging to the genera *Encrasicholina* or *Stolephorus*, or also in Honduras [88], the Philippines [89], the Azores [90], and Baja California (Mexico) [91]. The low amount of plankton ingested daily, on average, by the whale shark makes this ingestion route of little importance in the contribution to the total load of contamination assumed with the diet by the large filter feeder. To quantify the input of organochlorines through the ingestion of plankton based on the toxicological results obtained, we determined 74.60 g/day diet of plankton $\times$ 101.27 ng/g (HCB + DDTs + PCBs) in plankton = 7554.74 ng/day input of OCs with the plankton. Considering that a 4 m shark weighs about 500 kg [92], our findings would correspond to a daily intake of OCs equal to 15.11 ng/kg. This value turns out to be even lower than the quantity established for humans, which is defined as the daily allowable level without experiencing any harmful effects. The thresholds estimated by the World Health Organization (WHO) are 60 and 1200 µg/person/day, respectively, for PCBs and DDTs [93]. Considering a man who weighs 73 kg, these values correspond to an allowance of 822 ng/kg/day of PCBs and 16,438 ng/kg/day of DDTs. The EFSA Scientific Panel on Contaminants [94] estimated an average daily intake of non-dioxin like PCBs (NDL-PCBs) of about 15 ng/kg of body weight (assuming a body weight of 60 kg) per day for the "average" consumer, 20 ng/kg of body weight per day for large consumers of meat products, and about 35 ng/kg of body weight per day for large consumers of fish and fishery products. These comparisons were reported to demonstrate that a local whale shark's intake of these xenobiotics on a plankton-only diet is negligible. It would be interesting to have subcutaneous biopsies of these animals in order to evaluate the real levels of the accumulated organochlorines, and also to consider other routes of intake, such as small fishes, as has been suggested by Diamant et al. [86].

4. Conclusions

This study was the first to evaluate organochlorines in mesozooplankton along the island of Nosy Be that were sampled during the seasonal aggregation of whale sharks, and that were toxicologically evaluated by the function of the different feeding behaviors in this species. The results were preliminary, especially since the number of samples for each group was limited. Despite this, we can still conclude that the levels of the three xenobiotics investigated were lower than in other areas of the world. The fingerprint of DDTs is particularly interesting as it seems to highlight an important contamination by dicofol or, in any case, a recent introduction of DDT. Additionally, a greater presence of PCBs was observed compared to other xenobiotics, in particular of the congeners considered more recalcitrant, such as PCB153, PCB138, and PCB180. There were also no substantial differences between the feeding areas of the whale shark and those where the whale shark was not encountered, either in terms of the characterization and the volume of biomass or the levels of contamination. Finally, we found that the amount of plankton consumed on average per day by the whale shark may not only be insufficient for the growth of its biomass and for its overall energy requirements but that it may also contribute a truly negligible amount of contaminants, and so it absolutely cannot be considered a potential toxicological hazard for this filter feeder shark.

Author Contributions: Conceptualization, L.M., G.C. and P.M.; data curation, L.M., G.C., P.B. and P.M.; formal analysis, L.M. and G.C.; investigation, L.M., G.C., P.R., R.R., P.B., F.R.R. and P.M.; methodology, L.M., G.C. and P.M.; project administration, L.M. and P.M.; resources, L.M. and P.M.; supervision, L.M. and P.M.; validation, L.M. and G.C.; writing—original draft, L.M., G.C. and P.M.; writing—review and editing, L.M., G.C. and P.M. All authors have read and agreed to the published version of the manuscript.

Funding: This research received no external funding.

Institutional Review Board Statement: The study did not require ethical approval.

Informed Consent Statement: Not applicable.

Data Availability Statement: Not applicable.

Acknowledgments: The authors are grateful to the staff of Manta Diving Center (Nosy Be) for the logistic support.

Conflicts of Interest: The authors declare no conflict of interest.

References

1. Bettinetti, R.; Manca, M. Understanding the Role of Zooplankton in Transfer of Pollutants through Trophic Food Webs. In *Zooplankton: Species Diversity, Distribution and Seasonal Dynamics*; Nova Science Publishers, Inc.: Hauppauge, NY, USA, 2013; p. 18.
2. Boldrocchi, G.; Moussa Omar, Y.; Rowat, D.; Bettinetti, R. First Results on Zooplankton Community Composition and Contamination by Some Persistent Organic Pollutants in the Gulf of Tadjoura (Djibouti). *Sci. Total Environ.* **2018**, *627*, 812–821. [CrossRef]
3. Nybakken, J.W. *Marine Biology: An Ecological Approach*, 5th ed.; Benjamin Cummings: San Francisco, CA, USA, 2001; ISBN 9780321030764.
4. Compagno, L.J.V. Bullhead, Mackerel and Carpet Sharks: Heterodontiformes, Lamniformes and Orectolobiformes. In *Sharks of the World: An Annotated and Illustrated Catalogue of Shark Species Known to Date*; Food and Agriculture Organization of the United Nations: Rome, Italy, 2001; ISBN 9789251045435.
5. Norman, B. Rhincodon typus. In *The IUCN Red List of Threatened Species 2000*; IUCN: Gland, Switzerland, 2000.
6. Norman, B. Rhincodon typus. In *The IUCN Red List of Threatened Species 2005*; IUCN: Gland, Switzerland, 2005.

7. Pierce, S.J.; Norman, B. Rhincodon typus. In *The IUCN Red List of Threatened Species 2016*; IUCN: Gland, Switzerland, 2016.
8. Colman, J.G. A Review of the Biology and Ecology of the Whale Shark. *J. Fish Biol.* **1997**, *51*, 1219–1234. [CrossRef]
9. Motta, P.J.; Maslanka, M.; Hueter, R.E.; Davis, R.L.; de la Parra, R.; Mulvany, S.L.; Habegger, M.L.; Strother, J.A.; Mara, K.R.; Gardiner, J.M.; et al. Feeding Anatomy, Filter-Feeding Rate, and Diet of Whale Sharks *Rhincodon Typus* during Surface Ram Filter Feeding off the Yucatan Peninsula, Mexico. *Zoology* **2010**, *113*, 199–212. [CrossRef]
10. Hacohen-Domené, A.; Galván-Magaña, F.; Ketchum-Mejia, J. Abundance of Whale Shark (*Rhincodon Typus*) Preferred Prey Species in the Southern Gulf of California, Mexico. *Cybium* **2006**, *30*, 99–102.
11. Reyes-Mendoza, O.; Cárdenas-Palomo, N.; Herrera-Silveira, J.; Mimila-Herrera, E.; Trujillo-Córdova, J.; Chiappa-Carrara, X.; Arceo-Carranza, D. Quantity and Quality of Prey Available for the Whale Shark (*Rhincodon typus*, Smith 1828) at the Mexican Caribbean Aggregation Site. *Reg. Stud. Mar. Sci.* **2021**, *43*, 101696. [CrossRef]
12. Bava, P.; Micarelli, P.; Buttino, I. Zooplankton Assemblage Diversity in the Whale Shark *Rhincodon Typus* Aggregation Area of Nosy Be (Madagascar). *Estuar. Coast. Shelf Sci.* **2022**, *279*, 108159. [CrossRef]
13. Petatán-Ramírez, D.; Whitehead, D.A.; Guerrero-Izquierdo, T.; Ojeda-Ruiz, M.A.; Becerril-García, E.E. Habitat Suitability of *Rhincodon Typus* in Three Localities of the Gulf of California: Environmental Drivers of Seasonal Aggregations. *J. Fish Biol.* **2020**, *97*, 1177–1186. [CrossRef]
14. Taylor, L.R.; Compagno, L.J.V.; Struhsaker, P.J. Megamouth—A New Species, Genus, and Family of Lamnoid Shark (*Megachasma pelagios*, Family Megachasmidae) from the Hawaiian Islands. *Proc. Calif. Acad. Sci.* **1983**, *43*, 87–110.
15. Rowat, D.; Brooks, K.; March, A.; McCarten, C.; Jouannet, D.; Riley, L.; Jeffreys, G.; Perri, M.; Vely, M.; Pardigon, B.; et al. Long-Term Membership of Whale Sharks (*Rhincodon typus*) in Coastal Aggregations in Seychelles and Djibouti. *Mar. Freshw. Res.* **2011**, *62*, 621–627. [CrossRef]
16. Sequeira, A.M.M.; Mellin, C.; Fordham, D.A.; Meekan, M.G.; Bradshaw, C.J.A. Predicting Current and Future Global Distributions of Whale Sharks. *Glob. Change Biol.* **2014**, *20*, 778–789. [CrossRef]
17. Castro, A.L.F.; Stewart, B.S.; Wilson, S.G.; Hueter, R.E.; Meekan, M.G.; Motta, P.J.; Bowen, B.W.; Karl, S.A. Population Genetic Structure of Earth's Largest Fish, the Whale Shark (*Rhincodon typus*). *Mol. Ecol.* **2007**, *16*, 5183–5192. [CrossRef]
18. Pierce, S.J.; Méndez-Jiménez, A.; Collins, K.; Rosero-Caicedo, M.; Monadjem, A. Developing a Code of Conduct for Whale Shark Interactions in Mozambique. *Aquat. Conserv. Mar. Freshw. Ecosyst.* **2010**, *20*, 782–788. [CrossRef]
19. Berumen, M.L.; Braun, C.D.; Cochran, J.E.M.; Skomal, G.B.; Thorrold, S.R. Movement Patterns of Juvenile Whale Sharks Tagged at an Aggregation Site in the Red Sea. *PLoS ONE* **2014**, *9*, e103536. [CrossRef]
20. Fossi, M.C.; Baini, M.; Panti, C.; Galli, M.; Jiménez, B.; Muñoz-Arnanz, J.; Marsili, L.; Finoia, M.G.; Ramírez-Macías, D. Are Whale Sharks Exposed to Persistent Organic Pollutants and Plastic Pollution in the Gulf of California (Mexico)? First Ecotoxicological Investigation Using Skin Biopsies. *Comp. Biochem. Physiol. Part C Toxicol. Pharmacol.* **2017**, *199*, 48–58. [CrossRef]
21. Heyman, W.D.; Graham, R.T.; Kjerfve, B.; Johannes, R.E. Whale Sharks *Rhincodon Typus* Aggregate to Feed on Fish Spawn in Belize. *Mar. Ecol. Prog. Ser.* **2001**, *215*, 275–282. [CrossRef]
22. Meekan, M.G.; Jarman, S.N.; McLean, C.; Schultz, M.B.; Meekan, M.G.; Jarman, S.N.; McLean, C.; Schultz, M.B. DNA Evidence of Whale Sharks (*Rhincodon typus*) Feeding on Red Crab (*Gecarcoidea natalis*) Larvae at Christmas Island, Australia. *Mar. Freshw. Res.* **2009**, *60*, 607–609. [CrossRef]
23. Meekan, M.G.; Bradshaw, C.J.A.; Press, M.; McLean, C.; Richards, A.; Quasnichka, S.; Taylor, J.G. Population Size and Structure of Whale Sharks *Rhincodon typus* at Ningaloo Reef, Western Australia. *Mar. Ecol. Prog. Ser.* **2006**, *319*, 275–285. [CrossRef]
24. Boldrocchi, G.; Monticelli, D.; Butti, L.; Omar, M.; Bettinetti, R. First Concurrent Assessment of Elemental- and Organic-Contaminant Loads in Skin Biopsies of Whale Sharks from Djibouti. *Sci. Total Environ.* **2020**, *722*, 137841. [CrossRef]
25. Davis, D.; Banks, S.; Birtles, A.; Valentine, P.; Cuthill, M. Whale Sharks in Ningaloo Marine Park: Managing Tourism in an Australian Marine Protected Area. *Tour. Manag.* **1997**, *18*, 259–271. [CrossRef]
26. Graham, R.; Irvine, T.R.; Keesing, J.K. Iterative Planning and Adaptive Management of Whale Shark Tourism in Belize: Global Implications of Lessons Learnt from 1998 and 2004. In *CSIRO Marine and Atmospheric Research*; CSIRO: Canberra, Australia, 2007; pp. 73–74.
27. Quiros, A.L. Tourist Compliance to a Code of Conduct and the Resulting Effects on Whale Shark (*Rhincodon typus*) Behavior in Donsol, Philippines. *Fish. Res.* **2007**, *84*, 102–108. [CrossRef]
28. Rowat, D.; Meekan, M.G.; Engelhardt, U.; Pardigon, B.; Vely, M. Aggregations of Juvenile Whale Sharks (*Rhincodon typus*) in the Gulf of Tadjoura, Djibouti. *Environ. Biol. Fishes* **2007**, *80*, 465–472. [CrossRef]
29. Fossi, M.C.; Coppola, D.; Baini, M.; Giannetti, M.; Guerranti, C.; Marsili, L.; Panti, C.; de Sabata, E.; Clò, S. Large Filter Feeding Marine Organisms as Indicators of Microplastic in the Pelagic Environment: The Case Studies of the Mediterranean Basking Shark (*Cetorhinus maximus*) and Fin Whale (*Balaenoptera physalus*). *Mar. Environ. Res.* **2014**, *100*, 17–24. [CrossRef]
30. Tiktak, G.P.; Butcher, D.; Lawrence, P.J.; Norrey, J.; Bradley, L.; Shaw, K.; Preziosi, R.; Megson, D. Are Concentrations of Pollutants in Sharks, Rays and Skates (Elasmobranchii) a Cause for Concern? A Systematic Review. *Mar. Pollut. Bull.* **2020**, *160*, 111701. [CrossRef]
31. Consales, G.; Marsili, L. Assessment of the Conservation Status of Chondrichthyans: Underestimation of the Pollution Threat. *Eur. Zool. J.* **2021**, *87*, 165–180. [CrossRef]
32. Chiuchiolo, A.L.; Dickhut, R.M.; Cochran, M.A.; Ducklow, H.W. Persistent Organic Pollutants at the Base of the Antarctic Marine Food Web. *Environ. Sci. Technol.* **2004**, *38*, 3551–3557. [CrossRef]

33. Castro-Jiménez, J.; Bănaru, D.; Chen, C.-T.; Jiménez, B.; Muñoz-Arnanz, J.; Deviller, G.; Sempéré, R. Persistent Organic Pollutants Burden, Trophic Magnification and Risk in a Pelagic Food Web from Coastal NW Mediterranean Sea. *Environ. Sci. Technol.* **2021**, *55*, 9557–9568. [CrossRef]
34. Mäkinen, K.; Elfving, M.; Hänninen, J.; Laaksonen, L.; Rajasilta, M.; Vuorinen, I.; Suomela, J.-P. Fatty Acid Composition and Lipid Content in the Copepod Limnocalanus Macrurus during Summer in the Southern Bothnian Sea. *Helgol. Mar. Res.* **2017**, *71*, 11. [CrossRef]
35. Harmelin-Vivien, M.; Bănaru, D.; Dromard, C.R.; Ourgaud, M.; Carlotti, F. Biochemical Composition and Energy Content of Size-Fractionated Zooplankton East of the Kerguelen Islands. *Polar Biol.* **2019**, *42*, 603–617. [CrossRef]
36. Basu, S.; Chanda, A.; Gogoi, P.; Bhattacharyya, S. Organochlorine Pesticides and Heavy Metals in the Zooplankton, Fishes, and Shrimps of Tropical Shallow Tidal Creeks and the Associated Human Health Risk. *Mar. Pollut. Bull.* **2021**, *165*, 112170. [CrossRef]
37. UNEP. Report of the Persistent Organic Pollutants Review Committee on the Work of Its Seventeenth Meeting. In Proceedings of the Seventeenth Meeting of the Persistent Organic Pollutants Review Committee, Geneva, Switzerland, 24–28 January 2022; p. 56.
38. Marsili, L.; Jiménez, B.; Borrell, A. Chapter 7—Persistent Organic Pollutants in Cetaceans Living in a Hotspot Area: The Mediterranean Sea. In *Marine Mammal Ecotoxicology*; Fossi, M.C., Panti, C., Eds.; Academic Press: Cambridge, MA, USA, 2018; pp. 185–212. ISBN 978-0-12-812144-3.
39. Marsili, L.; Coppola, D.; Giannetti, M.; Casini, S.; Fossi, M.C.; van Wyk, J.; Sperone, E.; Tripepi, S.; Micarelli, P.; Rizzuto, S. Skin Biopsies as a Sensitive Non-Lethal Technique for the Ecotoxicological Studies of Great White Shark (*Carcharodon carcharias*) Sampled in South Africa. *Expert Opin. Environ. Biol.* **2016**, *4*, 1000126. [CrossRef]
40. Maddocks, R.F. Distribution Patterns of Living and Subfossil Podocopid Ostracodes in the Nosy Bé Area, Northern Madagascar. *Univ. Kans. Paleontol. Contrib.* **1966**, *72*, 12.
41. Goodman, S.M.; Benstead, J.P. (Eds.) *The Natural History of Madagascar*; University of Chicago Press: Chicago, IL, USA, 2003; ISBN 9780226303062.
42. Jonahson, M.; Harding, S. Occurrence of Whale Sharks (*Rhincodon typus*) in Madagascar. *Fish. Res.* **2007**, *84*, 132–135. [CrossRef]
43. Kiszka, J.; van der Elst, R. 11. Elasmobranchs (Sharks and Rays): A Review of Status, Distribution and Interaction with Fisheries of the Southwest Indian Ocean. In *Offshore Fisheries of the Southwest Indian Ocean*; Oceanographic Research Institute: Durban, South Africa, 2015; pp. 365–390.
44. Brenier, A.; Vogel, A. Integrating Conservation and Development in Madagascar's Marine Protected Areas. In *Marine Protected Areas: Interactions with Fishery Livelihoods and Food Security*; FAO: Rome, Italy, 2017; ISBN 9789251096062.
45. Diamant, S.; Rohner, C.A.; Kiszka, J.J.; d'Echon, A.G.; d'Echon, T.G.; Sourisseau, E.; Pierce, S.J. Movements and Habitat Use of Satellite-Tagged Whale Sharks off Western Madagascar. *Endanger. Species Res.* **2018**, *36*, 49–58. [CrossRef]
46. Rohner, C.A.; Richardson, A.J.; Prebble, C.E.M.; Marshall, A.D.; Bennett, M.B.; Weeks, S.J.; Cliff, G.; Wintner, S.P.; Pierce, S.J. Laser Photogrammetry Improves Size and Demographic Estimates for Whale Sharks. *PeerJ* **2015**, *3*, e886. [CrossRef]
47. Nelson, J.D.; Eckert, S.A. Foraging Ecology of Whale Sharks (*Rhincodon typus*) within Bahía de Los Angeles, Baja California Norte, México. *Fish. Res.* **2007**, *84*, 47–64. [CrossRef]
48. Giam, C.S.; Wong, M.K.; Hanks, A.R.; Sackett, W.M.; Richardson, R.L. Chlorinated Hydrocarbons in Plankton from the Gulf of Mexico and Northern Caribbean. *Bull. Environ. Contam. Toxicol.* **1973**, *9*, 376–382. [CrossRef]
49. Linko, R.R.; Raniamaki, P.; Urpo, K. PCB Residues in Plankton and Sediment in the Southwestern Coast of Finland. *Bull. Environ. Contam. Toxicol.* **1974**, *12*, 733–738. [CrossRef]
50. Joiris, C.R.; Overloop, W. PCBs and Organochlorine Pesticides in Phytoplankton and Zooplankton in the Indian Sector of the Southern Ocean. *Antarct. Sci.* **1991**, *3*, 371–377. [CrossRef]
51. Focardi, S.; Bargagli, R.; Corsolini, S. Organochlorines in Antarctic Marine Food Chain at Terranova Bay (Ross Sea). *Korean J. Polar Res.* **1993**, *4*, 73–77.
52. Ray, S.; Paranjape, M.A.; Koenig, B.; Paterson, G.; Metcalfe, T.; Metcalfe, C. Polychlorinated Biphenyls and Other Organochlorine Compounds in Marine Zooplankton off the East Coast of Newfoundland, Canada. *Mar. Environ. Res.* **1999**, *47*, 103–116. [CrossRef]
53. Fossi, M.C.; Borsani, J.F.; Di Mento, R.; Marsili, L.; Casini, S.; Neri, G.; Mori, G.; Ancora, S.; Leonzio, C.; Minutoli, R.; et al. Multi-Trial Biomarker Approach in Meganyctiphanes Norvegica: A Potential Early Indicator of Health Status of the Mediterranean "Whale Sanctuary". *Mar. Environ. Res.* **2002**, *54*, 761–767. [CrossRef]
54. Quental, T.; Ferreira, A.M.; Vale, C. The Distribution of PCBs and DDTs in Seston and Plankton along the Portuguese Coast. *Acta Oecol.* **2003**, *24*, S333–S339. [CrossRef]
55. Berrojalbiz, N.; Dachs, J.; Ojeda, M.J.; Valle, M.C.; Castro-Jiménez, J.; Wollgast, J.; Ghiani, M.; Hanke, G.; Zaldivar, J.M. Biogeochemical and Physical Controls on Concentrations of Polycyclic Aromatic Hydrocarbons in Water and Plankton of the Mediterranean and Black Seas. *Glob. Biogeochem. Cycles* **2011**, *25*, GB4003. [CrossRef]
56. Frouin, H.; Dangerfield, N.; Macdonald, R.W.; Galbraith, M.; Crewe, N.; Shaw, P.; Mackas, D.; Ross, P.S. Partitioning and Bioaccumulation of PCBs and PBDEs in Marine Plankton from the Strait of Georgia, British Columbia, Canada. *Prog. Oceanogr.* **2013**, *115*, 65–75. [CrossRef]
57. Desforges, J.-P.W.; Dangerfield, N.; Shaw, P.D.; Ross, P.S. Heightened Biological Uptake of Polybrominated Diphenyl Ethers Relative to Polychlorinated Biphenyls Near-Source Revealed by Sediment and Plankton Profiles along a Coastal Transect in British Columbia. *Environ. Sci. Technol.* **2014**, *48*, 6981–6988. [CrossRef]

58. Morales, L.; Dachs, J.; Fernández-Pinos, M.-C.; Berrojalbiz, N.; Mompean, C.; González-Gaya, B.; Jiménez, B.; Bode, A.; Ábalos, M.; Abad, E. Oceanic Sink and Biogeochemical Controls on the Accumulation of Polychlorinated Dibenzo-p-Dioxins, Dibenzofurans, and Biphenyls in Plankton. *Environ. Sci. Technol.* **2015**, *49*, 13853–13861. [CrossRef]
59. Kang, Y.; Zhang, R.; Yu, K.; Han, M.; Pei, J.; Chen, Z.; Wang, Y. Organochlorine Pesticides (OCPs) in Corals and Plankton from a Coastal Coral Reef Ecosystem, South China Sea. *Environ. Res.* **2022**, *214*, 114060. [CrossRef]
60. Wolff, M.S.; Thornton, J.; Fischbein, A.; Lilis, R.; Selikoff, I.J. Disposition of Polychlorinated Biphenyl Congeners in Occupationally Exposed Persons. *Toxicol. Appl. Pharmacol.* **1982**, *62*, 294–306. [CrossRef]
61. Safe, S.; Bandiera, S.; Sawyer, T.; Zmudzka, B.; Mason, G.; Romkes, M.; Denomme, M.A.; Sparling, J.; Okey, A.B.; Fujita, T. Effects of Structure on Binding to the 2,3,7,8-TCDD Receptor Protein and AHH Induction—Halogenated Biphenyls. *Environ. Health Perspect.* **1985**, *61*, 21–33. [CrossRef]
62. Bush, B.; Snow, J.; Koblintz, R. Polychlorobiphenyl (PCB) Congeners,p,P′-DDE, and Hexachlorobenzene in Maternal and Fetal Cord Blood from Mothers in Upstate New York. *Arch. Environ. Contam. Toxicol.* **1984**, *13*, 517–527. [CrossRef]
63. Clarke, J.U. Structure-Activity Relationships in PCBs: Use of Principal Components Analysis to Predict Inducers of Mixed-Function Oxidase Activity. *Chemosphere* **1986**, *15*, 275–287. [CrossRef]
64. Marsili, L.; D'Agostino, A.; Bucalossi, D.; Malatesta, T.; Fossi, M.C. Theoretical Models to Evaluate Hazard Due to Organochlorine Compounds (OCs) in Mediterranean Striped Dolphin (*Stenella coeruleoalba*). *Chemosphere* **2004**, *56*, 791–801. [CrossRef]
65. World Health Organization (WHO) (Ed.) *DDT and Its Derivatives: Environmental Aspects*; World Health Organization: Geneva, Switzerland, 1989; ISBN 9789241542838.
66. Aguilar, A. Relationship of DDE/ΣDDT in Marine Mammals to the Chronology of DDT Input into the Ecosystem. *Can. J. Fish. Aquat. Sci.* **1984**, *41*, 840–844. [CrossRef]
67. Borrell, A.; Aguilar, A. Variations in DDE Percentage Correlated with Total DDT Burden in the Blubber of Fin and Sei Whales. *Mar. Pollut. Bull.* **1987**, *18*, 70–74. [CrossRef]
68. Tsydenova, O.; Minh, T.B.; Kajiwara, N.; Batoev, V.; Tanabe, S. Recent Contamination by Persistent Organochlorines in Baikal Seal (Phoca Sibirica) from Lake Baikal, Russia. *Mar. Pollut. Bull.* **2004**, *48*, 749–758. [CrossRef]
69. Qiu, X.; Zhu, T.; Li, J.; Pan, H.; Li, Q.; Miao, G.; Gong, J. Organochlorine Pesticides in the Air around the Taihu Lake, China. *Environ. Sci. Technol.* **2004**, *38*, 1368–1374. [CrossRef]
70. Muñoz-Arnanz, J.; Jiménez, B. New DDT Inputs after 30 Years of Prohibition in Spain. A Case Study in Agricultural Soils from South-Western Spain. *Environ. Pollut.* **2011**, *159*, 3640–3646. [CrossRef]
71. Nowell, L.H.; Capel, P.D.; Dileanis, P.D. *Pesticides in Stream Sediment and Aquatic Biota: Distribution, Trends, and Governing Factors*; Pesticides in the Hydrologic System; Lewis Publishers: Boca Raton, FL, USA, 1999; ISBN 9781566704694.
72. Viselli, R. *La Contaminazione da DDT e Dai Suoi Derivati Nei Comparti Abiotici del Bacino del Basso Toce e del Lago Maggiore*; Istituto Superiore per la Protezione e la Ricerca Ambientale: Rome, Italy, 1999.
73. Qiu, X.; Zhu, T.; Yao, B.; Hu, J.; Hu, S. Contribution of Dicofol to the Current DDT Pollution in China. *Environ. Sci. Technol.* **2005**, *39*, 4385–4390. [CrossRef]
74. Qiu, X.; Zhu, T. Using the o,P′-DDT/p,P′-DDT Ratio to Identify DDT Sources in China. *Chemosphere* **2010**, *81*, 1033–1038. [CrossRef]
75. Li, L.; Liu, J.; Hu, J. Global Inventory, Long-Range Transport and Environmental Distribution of Dicofol. *Environ. Sci. Technol.* **2015**, *49*, 212–222. [CrossRef]
76. Beasley, V.R. Direct and Indirect Effects of Environmental Contaminants on Amphibians. In *Reference Module in Earth Systems and Environmental Sciences*; Elsevier: Amsterdam, The Netherlands, 2020; ISBN 9780124095489.
77. European Commission. Commission Delegated Regulation (EU) 2022/643 of 10 February 2022 Amending Regulation (EU) No 649/2012 of the European Parliament and of the Council as Regards the Listing of Pesticides, Industrial Chemicals, Persistent Organic Pollutants and Mercury and an Update of Customs Codes (Text with EEA Relevance). *Off. J. Eur. Union* **2022**, *118*, 14–54.
78. Turgut, C.; Gokbulut, C.; Cutright, T.J. Contents and Sources of DDT Impurities in Dicofol Formulations in Turkey. *Environ. Sci. Pollut. Res.* **2009**, *16*, 214–217. [CrossRef]
79. Global Environment Facility—GEF. *Improvement of DDT-Based Production of Dicofol and Introduction of Alternative Technologies Including IPM for Leaf Mites Control in China*; Global Environment Facility—GEF: Washington, DC, USA, 2006.
80. Wong, F.; Alegria, H.A.; Bidleman, T.F.; Alvarado, V.; Angeles, F.; Galarza, A.Á.; Bandala, E.R.; de la Cerda Hinojosa, I.; Estrada, I.G.; Reyes, G.G.; et al. Passive Air Sampling of Organochlorine Pesticides in Mexico. *Environ. Sci. Technol.* **2009**, *43*, 704–710. [CrossRef]
81. Li, J.; Zhang, G.; Qi, S.; Li, X.; Peng, X. Concentrations, Enantiomeric Compositions, and Sources of HCH, DDT and Chlordane in Soils from the Pearl River Delta, South China. *Sci. Total Environ.* **2006**, *372*, 215–224. [CrossRef]
82. Aguilar, A.; Borrell, A.; Pastor, T. Biological Factors Affecting Variability of Persistent Pollutant Levels in Cetaceans. *J. Cetacean Res. Manag.* **1999**, 83–116. [CrossRef]
83. United States Department of State. *Madagascar*; Background Notes Series; Bureau of Public Affairs: Washington, DC, USA, 1987; pp. 1–7.
84. Jenkins, C.N.; Van Houtan, K.S. Global and Regional Priorities for Marine Biodiversity Protection. *Biol. Conserv.* **2016**, *204*, 333–339. [CrossRef]

85. Di Capua, I.; Micarelli, P.; Tempesti, J.; Reinero, F.R.; Buttino, I. Zooplankton Size Structure in the Gulf of Tadjoura (Djibouti) during Whale Shark Sighting: A Preliminary Study. *Cah. Biol. Mar.* **2021**, *62*, 290–294. [CrossRef]
86. Diamant, S.; Pierce, S.J.; Rohner, C.A.; Graham, R.T.; Guillemain d'Echon, A.; Guillemain d'Echon, T.; Sourisseau, E.; Fidiarisandratra, L.C.; Bakary, G.; Trélanche, S.; et al. Population Structure, Residency, and Abundance of Whale Sharks in the Coastal Waters off Nosy Be, North-Western Madagascar. *Aquat. Conserv. Mar. Freshw. Ecosyst.* **2021**, *31*, 3492–3506. [CrossRef]
87. Boldrocchi, G.; Bettinetti, R. Whale Shark Foraging on Baitfish off Djibouti. *Mar. Biodivers.* **2019**, *49*, 2013–2016. [CrossRef]
88. Fox, S.; Foisy, I.; De La Parra Venegas, R.; Galván Pastoriza, B.E.; Graham, R.T.; Hoffmayer, E.R.; Holmberg, J.; Pierce, S.J. Population Structure and Residency of Whale Sharks *Rhincodon Typus* at Utila, Bay Islands, Honduras. *J. Fish Biol.* **2013**, *83*, 574–587. [CrossRef]
89. Araujo, G.; Agustines, A.; Tracey, B.; Snow, S.; Labaja, J.; Ponzo, A. Photo-ID and Telemetry Highlight a Global Whale Shark Hotspot in Palawan, Philippines. *Sci. Rep.* **2019**, *9*, 17209. [CrossRef] [PubMed]
90. Fontes, J.; McGinty, N.; Machete, M.; Afonso, P. Whale Shark-Tuna Associations, Insights from a Small Pole-and-Line Fishery from the Mid-North Atlantic. *Fish. Res.* **2020**, *229*, 105598. [CrossRef]
91. Montero-Quintana, A.N.; Ocampo-Valdez, C.F.; Vázquez-Haikin, J.A.; Sosa-Nishizaki, O.; Osorio-Beristain, M. Correction to: Whale Shark (*Rhincodon Typus*) Predatory Flexible Feeding Behaviors on Schooling Fish. *J. Ethol.* **2022**, *40*, 117. [CrossRef]
92. Matsumoto, R.; Toda, M.; Matsumoto, Y.; Ueda, K.; Nakazato, M.; Sato, K.; Uchida, S. Chapter 2. Notes on Husbandry of Whale Sharks, *Rhincodon Typus*, in Aquaria. In *The Elasmobranch Husbandry Manual II: Recent Advances in the Care of Sharks, Rays and Their Relatives*; Ohio Biological Survey: Columbus, OH, USA, 2017; p. 504.
93. FAO; WHO. *Joint FAO/WHO Food Standards Programme: Codex Maximum Limits for Pesticides Residues*; FAO: Rome, Italy; WHO: Geneva, Switzerland, 1986.
94. EFSA. Opinion of the Scientific Panel on Contaminants in the Food Chain [CONTAM] Related to the Presence of Non Dioxin-like Polychlorinated Biphenyls (PCB) in Feed and Food. *EFSA J.* **2005**, *3*, 284. [CrossRef]

diversity

MDPI

Article

Unexpected Records of Newborn and Young Sharks in Ligurian and North Tyrrhenian Seas (North-Western Mediterranean Basin)

Cecilia Mancusi [1,2,*], Fabrizio Serena [3,*], Alessandra Neri [2,4], Umberto Scacco [5,6], Romano Teodosio Baino [1], Alessandro Voliani [7] and Letizia Marsili [2,8]

1 Environmental Protection Agency of Tuscany Region (ARPAT), Via G. Marradi 114, 57126 Livorno, Italy; rt.baino@arpat.toscana.it

2 Department of Physical Sciences, Earth and Environment, University of Siena, Via P. A. Mattioli 4, 53100 Siena, Italy; alessandra.neri@student.unisi.it (A.N.); marsilil@unisi.it (L.M.)

3 National Research Council—Institute of Marine Biological Resources and Biotechnology, Via Vaccara 61, 91026 Mazara del Vallo, Italy

4 CIBM—Consorzio per il Centro Interuniversitario di Biologia Marina ed Ecologia Applicata "G. Bacci", Viale N. Sauro 4, 57128 Livorno, Italy

5 Italian Institute for Environmental Protection and Research (ISPRA), National Centre of Laboratories-Biology, Via di Castel Romano 100, 00128 Rome, Italy; umberto.scacco@isprambiente.it or umbertoscacco@unitus.it

6 Department of Bio Ecological Sciences, University of Tuscia, Largo dell'Università snc, 01100 Viterbo, Italy

7 Independent Researcher, Viale Trieste 7, 57016 Rosignano Marittimo, Italy; a.voliani@gmail.com

8 CIRCE, Centro Interuniversitario di Ricerca sui Cetacei, Via P. A. Mattioli 4, 53100 Siena, Italy

* Correspondence: c.mancusi@arpat.toscana.it (C.M.); fabrizio.serena@irbim.cnr.it (F.S.)

Abstract: Between 2007 and 2022, 112 specimens of newborn and young pelagic sharks were recorded in the waters of Tuscany Region, in the South Ligurian–North Tyrrhenian Seas (north-western Mediterranean basin). The sharks belonged to the *Carcharhinus plumbeus* (n = 14), *Prionace glauca* (n = 66), *Isurus oxyrinchus* (n = 16), *Mobula mobular* (n = 5) *Alopias vulpinus* (n = 7) and *Hexanchus griseus* (n = 4) species. Each animal was correctly identified thanks to the photographs or videos collected. All specimens were incidentally captured with set nets in inshore shallow waters, except bluntnose six-gill sharks, which were bycatch of deep-water bottom-trawl fishery. Body mass, sex, total length and biometric measurements were recorded in 34 baby sharks following the Mediterranean Large Elasmobranches Monitoring (MEDLEM) protocol. The presence of very evident and often non-healed umbilical scar confirmed that some of the sample specimens were newborn. Further confirmation came from the comparison between the total length observed and the size at birth known for the sampled species as reported in the literature. Some baby sharks were preserved in the Museums of Natural History of Pisa and Florence University collections. The importance of the coastal area studied as a possible shark nursery is discussed.

Keywords: North Italian waters; young of the year; umbilical scar; incidental catch; MEDLEM

1. Introduction

The Mediterranean basin displays relatively high chondrichthyan richness, with 7% of the total number of elasmobranchs being represented inside the basin [1–7]. Due to their life history traits, sharks and rays are particularly susceptible to over-exploitation, and their populations have very low resilience. Species often show restricted distributions and small population sizes, dependent on mating, spawning, nursery and breeding grounds, or on specific habitats [4,8]. Although there is no real direct fishery targeting large cartilaginous fishes in the Mediterranean, they are incidentally caught, mainly with gillnets and bottom longlines targeting European hake [9]. Surface drifting longlines, targeting tuna and swordfish, also capture some pelagic shark species as bycatches or discards [9–15].

To face the biodiversity loss and to increase the effectiveness of the conservation measures in the Mediterranean basin, it is important to establish a common procedure to collect data on shark individuals that are accidentally captured, sighted at sea or stranded. In this light, the Mediterranean Large Elasmobranchs Monitoring (MEDLEM) database aims at contributing to the improvement in knowledge on the presence, spatial distribution and bycatch of large cartilaginous fishes species present in the Mediterranean and Black Seas. Officially established in 1985 [16], it became fully operational in 2000, beginning to record data on elasmobranch catches, sightings, strandings and historical records in a single database [17,18].

More recently, in Tuscany Region (Italy, north-western Mediterranean), monitoring activities foreseen by MEDLEM were included into the former Tuscany Observatory for Cetacean in 2007, now Tuscany Observatory for Biodiversity (Osservatorio Toscano Biodiversità (OTB)) sensu art. 11 Regional Law 30/2015, together with the monitoring of strandings and incidental capture of cetaceans and sea turtles. This action taken by Tuscany Region represents a real contribution to the international effort for the conservation of the marine ecosystem and its resources. In Tuscany, the Region Administration has created a coordinated and synergic system among its technical instruments, represented by ARPAT (Environmental Protection Agency, Tuscany Region), universities, research centers, museums, aquaria, environmental associations and fishermen [19].

The main features of the MEDLEM Database Application are (i) the implementation of data collection, especially for bycatch evaluation; (ii) the standardization of data entry procedures; (iii) effective data sharing among the participating countries; and (iv) free access to the website for participants.

In the Ligurian and North Tyrrhenian Seas, pelagic sharks represent a limited part of the commercial bycatch of professional longline fishery targeting sword fish and tuna. The most frequently caught species are *Prionace glauca* (Linneo, 1758), followed by *Isurus oxyrinchus* (Rafinesque, 1810), *Lamna nasus* (Bonnaterre, 1788), *Alopias vulpinus* (Bonnaterre, 1788) and *Carcharhinus plumbeus* (Nardo, 1827) [20]. Sharks of the genus *Carcharhinus* are represented in the Mediterranean Sea by at least eight species [21,22], but there is a great lack of information about species identity, abundance and distribution, mainly due to the scarcity of catches and the difficulty in their correct identification, especially for juveniles.

Due to the low reproduction rate of elasmobranchs, newborn records are very limited in comparison to other reproductive strategies, for example, those displayed by many osteichthyes. Knowledge about the early stages of juvenile sharks is extremely poor, and newborns require particular attention. Furthermore, the need to identify grounds of possible aggregation or nursery areas is very important for eventually establishing a protection area for a local and regional management plan of the maritime space.

Data gathered in the study area, shown in this paper, can provide useful information for this purpose. Juvenile numbers and detailed morphometrics, although representing a still-too-small data set for species and sex, feature an important initial database to be improved.

2. Materials and Methods

2.1. Study Area

The study area encompassed the marine waters of Tuscany, in Italy, between the Ligurian and the North Tyrrhenian Seas, in the north-western part of the Mediterranean basin; relating to bluntnose six-gill sharks records, we also considered the northernmost part of Latium Region (Figure 1). The data analyzed in this paper were collected in Tuscany during the period of 2007–2022.

2.2. Sampling Protocols

The sampling protocol adopted is characterized by the word "large", which defines the nature of the database (Mediterranean Large Elasmobranchs Monitoring). This refers to maximum size reached by each different species; particularly, we only consider sharks with more than 100 cm in total length (TL) and batoids with more than 150 cm in disc width (DW) as maximum sizes [23]. Applying this rule to the species in the Mediterranean and Black Seas, we restricted the MEDLEM protocol to 16 different families and 7 orders [24]. Thanks to the collaboration of several research institutes, military authorities, professional and recreational fishermen, and NGOs, a great amount of valuable information on catches, sightings and strandings of large cartilaginous fishes was archived into the regional database.

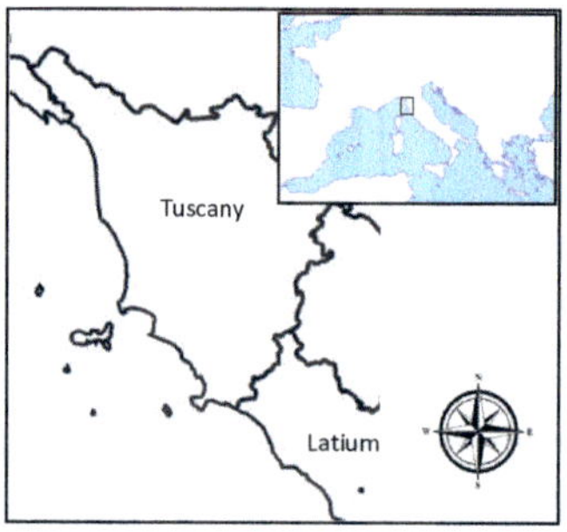

Figure 1. Map of the study area: coasts of Tuscany and northern Latium counties; Ligurian and North Tyrrhenian Seas (north-western Mediterranean basin).

When some of these stakeholders get involved in a capture event (or more rarely, in a stranding one), it is strongly recommended to immediately call the free blue number 1530, made available 24 h a day throughout the national territory by the Italian Port Authority; it is also very useful to take and send some photographs of the whole specimens or of some details (fins, teeth, ventral part, etc.), which can help correct taxonomic identification, sex attribution or size estimation. If the cartilaginous fish is still alive, it is advisable to free it at sea. On the contrary, if the specimens are dead, the Port Authority calls some reference numbers, in particular ARPAT's, as coordinator of the regional nets, which activate all the necessary procedures to recover the shark. Spatial–temporal data about catches, strandings or sightings (date, time, country, latitude, longitude, etc.) are registered; then, the species and fishing gear responsible for capture are annotated and communicated to the MEDLEM account (medlemcontact@gmail.com).

2.3. Biological and Biometrical Parameters

When possible, all biological parameters were collected. Total length was recorded to the nearest centimeter with a measurement tape; body mass, with an electronic dynamometer to the nearest gram; sex and all the other thirty-one biometrical measurements, following the MEDLEM protocol: fork length and from snout tip to pre-caudal pit, 1st dorsal origin, 2nd dorsal origin, pectoral fin origin, pelvic fin origin, anal fin origin, 1st gill opening and 5th gill opening; head (snout–eye length, snout–mouth length, snout–nostril length, 1st-5th gill opening length, horizontal diameter eye and vertical diameter eye); pectoral fin (base length, anterior margin length, posterior margin length and height); 1st dorsal fin (anterior margin length, posterior margin length, height and base length); caudal fin (dorsal lobe length, terminal margin length, sub-terminal margin length, ventral lobe length, post-ventral margin length and pre-ventral margin length); and claspers (outer length and inner length) [23].

Species identification and taxonomic nomenclature followed [22,25–27]. Some specimens were preserved in alcohol at 75% and were stored in the collections of Museum of Natural History of Pisa University and Museum of Natural History of Florence University, zoological section, La Specola. Photographs of the examined specimens were stored in the digital archives of the authors and are available for further comparisons. Genetic samples were collected from all the specimens.

For this paper's purpose, individuals were divided into four size groups based on the observed length and related information in the available literature: newborn, young of the year, juveniles and adults. Newborn (NB) size corresponded to the size at birth; young of the year (YOY) were considered those individuals of age 0; the separation between juveniles (JUV) and adults (ADL) was determined according to the total length (LT) at which 50% of the population reached sexual maturity (LT50). According to the literature, the four groups were established for six species of interest (Table 1).

Table 1. The four size groups established in the present paper according to the bibliography. NB = newborn; YOY = young of the year; JUV = juveniles; ADL = adult; TL = total length; DW = disc width.

Species	NB (cm)	YOY (cm)	JUV (cm)	ADL (cm)	References
I. oxyrinchus	$60 < TL \leq 70$	$70 < TL \leq 100$	$100 < TL \leq 200$	$TL > 200$	[28,29]
P. glauca	$35 < TL \leq 45$	$45 < TL \leq 80$	$81 < TL \leq 180$	$TL > 180$	[28,29]
C. plumbeus	$56 < TL \leq 75$	$75 < TL \leq 100$	$100 < TL \leq 140$	$TL > 140$	[29–31]
A. vulpinus	$114 < TL \leq 160$	$160 < TL \leq 170$	$170 < TL \leq 300$	$TL > 300$	[29,32,33]
H. griseus	$65 < TL \leq 74$	$74 < TL \leq 100$	$100 < TL \leq 300$	$TL > 300$	[29,34,35]
M. mobular	$160 < DW \leq 180$			$DW > 300$	[4,8,36]

We focused on newborn or young of the year specimens of six species: *P. glauca*, *I. oxyrinchus*, *A. vulpinus*, *C. plumbeus*, *H. griseus* and *M. mobular*.

2.4. Biometrical and Statistical Analyses

We show the average ± 1 SD, and minimum and maximum values of each measure for each species. In the case of a single specimen, we present one value. In the case of samples with size ≥ 3 and at least 3 records/sex, we tested if sexual differences in body morphometrics were present within each species (*P. glauca* and *C. plumbeus*) with Student's *t*-test and Levene's test for variance homogeneity. In addition, to detect if occurrence of sex was a stochastic or an actual pattern, we applied the χ^2 test with Yates's correction for <5 expected frequencies. We performed all analyses with IBM SPSS, 20.0 release.

3. Results

From 2007 to 2022, in the study area, 222 large elasmobranch records belonging to 10 different species were registered as bycatch, sightings or, rarely, stranding events; all the data of the findings are reported in Table S1 (Supplementary Materials). The most frequent species was the blue shark *P. glauca* (40 %) (Table 2; Figure 2).

Table 2. Elasmobranchs registered in South Ligurian–North Tyrrhenian Seas between 2007 and 2022; * = disc width.

Species	Total	Range (TL, cm)	Number of Measured Individuals
Prionace glauca	89	45–300	62
Mobula mobular	39	96–370 *	8
Hexanchus griseus	27	80–420	22
Alopias vulpinus	22	120–432	14
Isurus oxyrinchus	21	70–200	13
Carcharhinus plumbeus	15	54–211	11
Cetorhinus maximus	6	295–800	6
Alopias superciliosus	1		
Carcharodon carcharias	1	350	1
Aetomylaeus bovinus	1	80 *	1
Total	**222**		**138**

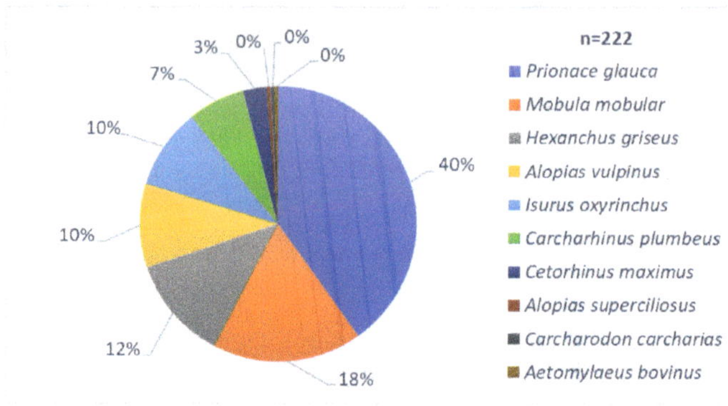

Figure 2. Percentage of species registered in the period 2007–2022 in the study area; 0% = 0.0045% (n = 1).

In total, 51% (n = 112) of the total elasmobranchs registered in the study area were "juveniles" (YUV + YOY + NB) and were exclusive to six species: *P. glauca, I. oxyrinchus, A. vulpinus, C. plumbeus, H. griseus* and *M. mobular*. A total of 78% of the data refer to bycatch events; 18%, to sightings; and only 4%, to strandings. Details of the fishing gear responsible for capture are shown in Figure 3.

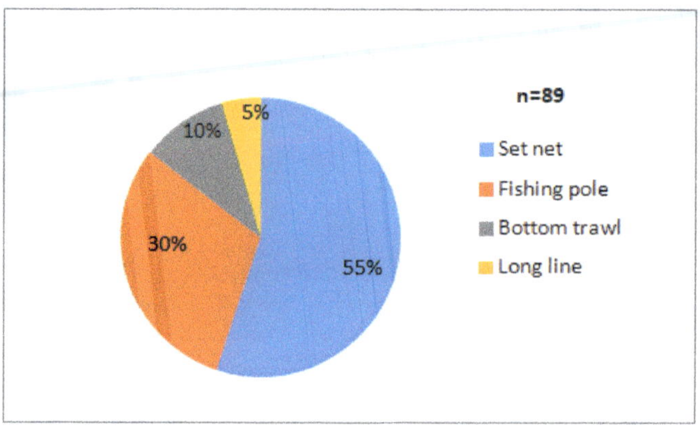

Figure 3. Types of fishing gear causing bycatch of young individuals in South Ligurian–North Tyrrhenian Seas.

Especially for three species, a predominance of juveniles or newborns with respect to adult specimens was highlighted as follows: *C. plumbeus* (93%), *I. oxyrinchus* (76%) and *P. glauca* (73%). When examined, some of these individuals showed an almost-healed umbilical scar visible on the ventral side, in the midpoint of the line joining the origin of the pectoral fins, to confirm their "newborn" condition; in particular, in seven sandbar sharks, the umbilical mark was still an open hole (on the left in Figure 4) [31].

For some species, mainly the three most abundant ones, juveniles and newborn individuals were concentrated in very restricted areas.

Detailed measurements of the 35 examined young individuals are reported in Table S2.

3.1. *Prionace Glauca*

Blue shark was the most abundant species, with 66 immature individuals: 26 JUV (90 < TL < 160 cm), 39 YOY (49,5 < TL < 81 cm) and 1 NB (TL = 45 cm). They were mainly captured with set nets or fishing poles, at depths between 3 and 75 m (exceptionally, one little blue shark was captured at 200 m with a bottom trawl), in two restricted areas close to the Meloria (Marine Protected Area) and Vada shoals (approximately 42.405–43.980° N) (Figure 5d); for the most part, sharks were captured or sighted in the spring–summer period (May to August) and were often captured alive and released at sea. Sixteen individuals were examined (Table S2).

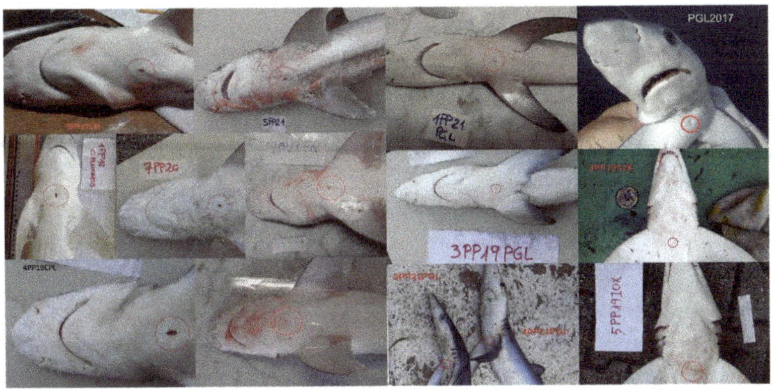

Figure 4. Some examples of the ventral side of the examined individuals where the almost-healed umbilical scar is more evident (red circle); PGL = blue shark *P. glauca*, CPL = sandbar shark *C. plumbeus* and IOX = mako shark *I. oxyrinchus*. Note the seven sandbar sharks (on the left) with the umbilical mark still open.

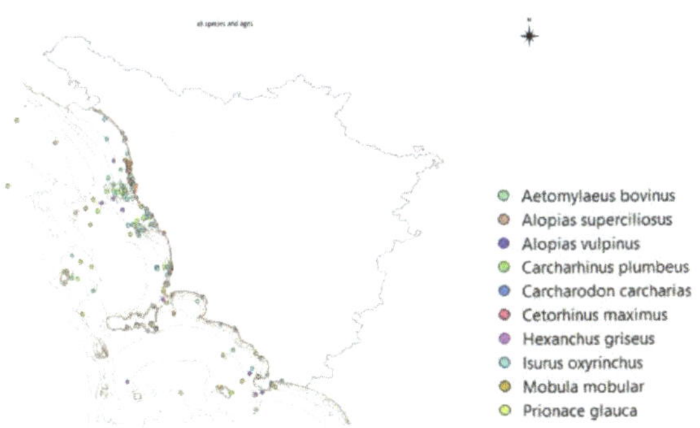

(a)

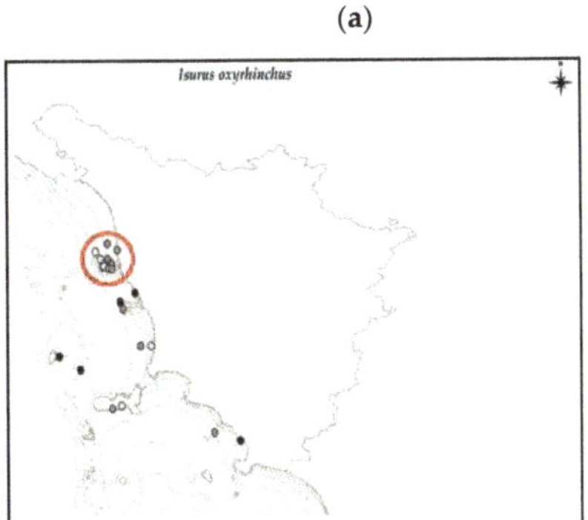

(b)

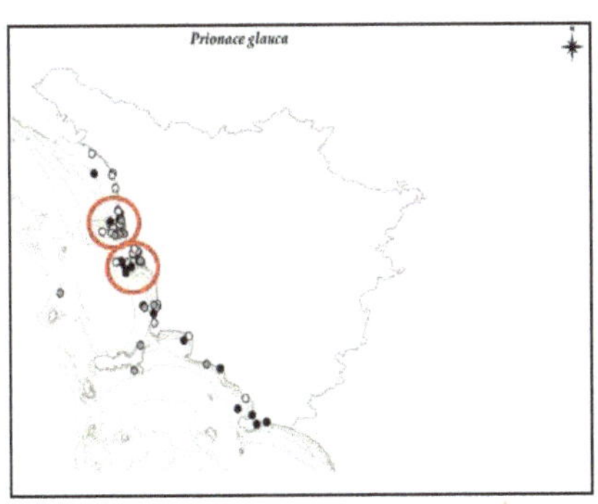

(c)

Figure 5. *Cont.*

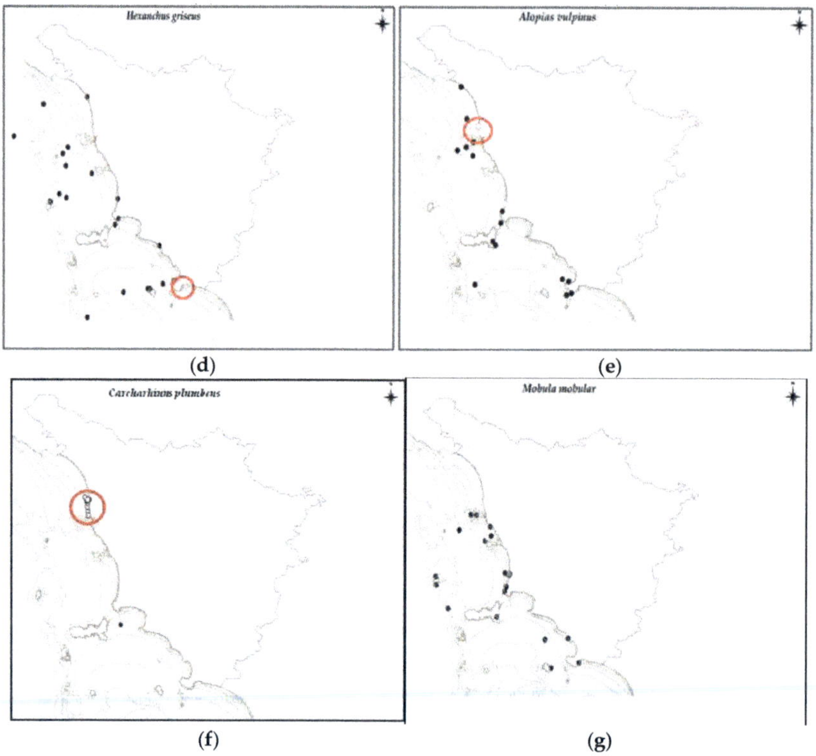

Figure 5. Geographical distribution of the findings, South Ligurian–North Tyrrhenian Seas (2007–2022). (**a**) All the species (222 individuals); (**b**) *I. oxyrinchus;* (**c**) *P. glauca;* (**d**) *H. griseus;* (**e**) *A. vulpinus;* (**f**) *C. plumbeus;* (**g**) *M. mobular*. Black symbols = adult individuals; white or gray symbols = juveniles (NB + YOY + JUV); red circles = coastal concentration of "juveniles".

3.2. *Carcharhinus Plumbeus*

A total of 14 newborn sharks were accidentally caught with set nets in recent years: 2018 (n = 1), 2019 (n = 1), 2020 (n = 2), 2021 (n = 9) and 2022 (n = 1). All the captures were registered in September and October and in a restricted area between Tirrenia and Gombo (Pisa) at the mouth of the Arno River and "Fiume Morto Nuovo" in very shallow waters (7–15 m depth) between 43.607° N and 43.736° N (Figure 5b). Total length ranged between 54 and 90 cm (n = 10); weight, between 0.9 and 4 kg (n = 9), in 2 females and 7 males. Four baby sharks were still alive and were released at sea. Seven dead sharks were examined; they had a still-open birth mark (Table S2, Figure 4).

3.3. *Isurus Oxyrinchus*

In the period 2016–2022, ten juveniles shortfin mako sharks were registered (106 cm < TL < 135 cm), together with five young of the year (80 < TL < 88.5 cm) and only one newborn of 70 cm in total length, released alive. These 16 "small" sharks were mainly captured between May and July with set nets and fishing poles in a restricted area between Livorno (Vada and Meloria shoals) and the mouth of the Arno River (Pisa), at approximately 42.601–43.678° N (Figure 5c) at depths of about 10 to 70 m. Six dead fish were examined; see Table S2 for details.

3.4. *Hexanchus Griseus*

Only four young individuals of this species were captured, two of them with bottom trawls at depths of 301–366 m and two in more shallow waters; two were considered juveniles, and two, young of the year. The YOY specimens were captured slightly to the south of the study area, in the northern part of Latium Region (North Tyrrhenian Sea) (Figure 5e); they were 80 and 87.5 cm long, a male and a female, respectively. Three dead fish were examined (Table S2).

3.5. *Alopias Vulpinus*

Seven individuals all classified as newborn were registered in 2017 (n = 2) and 2020 (n = 5). The total length ranged between 120 and 147 cm (n = 3); the weight, between 3.5 and 7 kg (n = 7). All the sharks except one were incidentally captured with set nets at depths between 7 and 15 m close to the Arno River (Pisa) at 43.628–43.677° N (Figure 5f). Two dead individuals were examined (Table S2).

3.6. *Mobula Mobular*

Five individuals were captured or sighted between 2008 and 2019 in coastal waters at a depth of about 20 m. Their disc width ranged between 96 and 180 cm. Only one ray was dead and was examined (Table S2bis).

The average $\pm$ 1 SD, and minimum and maximum values of each measure for each species are reported in Table S3. For the two most numerous species (see Biometrical and Statistical Analyses), we obtained the following data: For *P. glauca*, the average values of almost all body morphometrics were similar between males and females (all not significant, with p ranging from 0.65 to 0.866). It is worth noticing that two body measures slightly differed between sexes. In particular, the pectoral fin height resulted a little larger in females than in juvenile males ($p < 0.065$), as did the caudal fin sub-terminal margin length ($p < 0.096$). In *C. plumbeus*, the average values were identical or much similar between sexes (p ranging from 0.164 to 0.969), except the caudal fin ventral lobe length, which was lightly larger in females than males ($p < 0.03$). Detailed results (homogeneity tests, degrees of freedom and significance) of the comparison of body morphometrics between sexes for the two considered species are shown in Table S4.

The occurrence of sexes in our sample did not deviate significantly (χ^2 with Yate's correction = 0.518, 1 d.f., $p = 0.657$).

4. Discussion

Prionace glauca is a widespread species in the Mediterranean, and it often represents the most important bycatch fraction of tuna and swordfish longline fishery, especially in Italy, Malta, Morocco and Tunisia [9]. This species is also incidentally captured in recreational fishing in Tuscany. It is listed in Annex 3 of the Berna and Barcelona Conventions and as Vulnerable in the IUCN Red List for the Mediterranean basin. Adults reach 380 cm in total length (TL), generally between 180 and 300 cm; the total length of newborns, usually 15–30 in number, is 35–45 cm [23,37,38]. The observation of newborns in the period of May–September is compatible with the fact that complete embryos have been observed by other authors in May–July [37,39,40].

Isurus oxyrinchus is occasionally caught with swordfish longlines. Its very low reproductive capacity may cause a rapid decline also in the Mediterranean basin and Italian seas [41]. For this reason, shortfin mako shark is listed in Annex A3 of the Berna Convention and Annex A2 of the Bonn and Berna Conventions; moreover, it is listed as Critically Endangered in the IUCN Red List. Adults reach 400 cm in total length (TL), generally 150–200 cm in TL, while newborns, usually 4–25 in number, are 60–70 cm in TL [23,36].

The cosmopolitan species, i.e., present in all the Mediterranean basin, *A. vulpinus* is often bycatch in professional fishery. Listed in Annex 3 of the Barcelona Convention and as Vulnerable in the IUCN Red List, adults reach 600 cm. At birth, young specimens are more than 100 cm long, up to 120 cm in TL [29,32,33].

Carcharhinus plumbeus is an endangered shark species in the Mediterranean Sea [42] considered a protected species in the inventory of Turkish Fisheries Act Marine Protected Species [43]. It is listed in Annex 3 of the Barcelona Convention and as Endangered in the IUCN Red List. Adults can reach 240–300 cm in total length (TL), generally 220 cm in TL, while newborns are 45–75 cm in TL, usually 5–12 in number [23,36]. In the Mediterranean Sea, the size at birth of *C. plumbeus* ranges from 45 cm [44,45] to 65 cm [44–46]. The species grows larger in western Atlantic waters compared with Mediterranean waters, as the maximum size at birth recorded was 72 cm [30], and it was 75 cm for the overall northern Atlantic [47]. Our sample ranged between 54 and 90 cm in TL, which is a relatively larger length interval than the data cited above. According to Capapé [46], in the waters of Gulf Gabès, the highest number of sandbar shark juveniles was observed in the summer months. According to Carlsson [30], neonate sandbar sharks (<age 1) usually reside in primary nursery areas, where they are born in the first period of summer. Therefore, it can be assumed that the specimens described in the present paper either (1) were born and resided off the Tirrenia–Gombo coast (Pisa) or (2) were born elsewhere and migrated from another area. At present, the available data do not allow us to give a definitive answer to these questions. As a matter of fact, the occurrence of several specimens (n = 7) with a partially healed umbilical scar (a still-"open" mark) could indicate that the coast of Pisa can serve as a probable nursery ground for *C. plumbeus*. This is a promising possibility for the survival of such an endangered shark species in the Mediterranean Sea. Similar considerations can also be valid for other species, such as *I. oxyrinchus* and *P. glauca*.

Recently, in the current year (21 May 2023), all *Carcharinidi* spp. entered Appendix II of Cites (https://cites.org/eng/app/appendices.php (accessed on 21 May 2023)).

On one hand, the ecological characteristics of the coastal waters of northern Tuscany can indicate a potential favorable habitat for YOY individuals of different large elasmobranch species, according to the conventional shark nursery area theory (e.g., having high productivity and low exposure to potential predators; and adults and juveniles in open and coastal waters, respectively) [48].

On the other hand, the high abundance of both YOY and juvenile sharks alone does not warrant a definitive classification of the study area as a nursery. According to Heupel et al. [49], three testable criteria should be examined for an area to be considered a shark nursery: (1) higher-than-average density of YOY sharks in the area, (2) tendency for YOY sharks to remain or return to the area for extended periods and (3) the area being used repeatedly over the years. Further effort will be spent to verify these criteria; particularly, there is a need for accurate collection of fishery data and the implementation of tagging studies in the study area, as well as the monitoring of environmental conditions, especially related to water temperature, which appears to play an important role in defining the putative nursery habitat of YOY blue and shortfin mako sharks [28].

Regarding body size analysis, the investigated sample is undoubtedly too small to appropriately represent the morphological variability of each species. However, some weak differences that we detected in three body morphometrics suggest that sexual size dimorphism could be exhibited even in the newborn or juvenile stage in the investigated species. Forthcoming studies on larger samples per species and per sex will be necessary to address the presence or absence of early sexual dimorphism in body measures.

The apparent growing trend of reports of large cartilaginous fishes in our region, especially of "juveniles" (Figure 6), does not correspond to a real numerical increase in these species but rather to ever-increasing attention paid to and sensitivity towards the problems related to the exploitation and conservation of this group of fishes. From the foregoing, it emerges that our region is confirmed as an area of high interest from the marine biodiversity point of view, where it is necessary to perform monitoring activity with ever-increasing commitment.

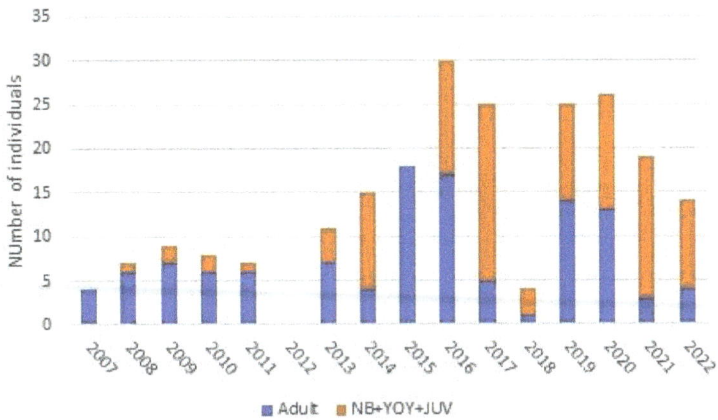

Figure 6. Trend of the individuals (adult + "juveniles") recorded over the years in the study area.

For some species, the phenomenon we are observing suggests a probable recovery in terms of abundance. In particular, in Italian waters, in the last decade, we observed a positive trend referred, above all, to the presence of young specimens of blue shark and shortfin mako in the bycatch of small-scale fishery [50].

Supplementary Materials: The following supporting information can be downloaded at: https://www.mdpi.com/article/10.3390/d15070806/s1, Table S1: All the detailed information about the findings; LAT = latitude, LONG = longitude, ADL = adult, JUV = juvenile, NB = newborn, Table S2: Detailed measurements of 35 examined young individuals, Table S2bis: Detailed measurements of young *M. mobular* examined, Table S3: Descriptive statistics (average ± 1 SD, and minimum and maximum values) of each measure for each species, Table S4: Detailed results (homogeneity tests, degrees of freedom and significance) of the comparison of body morphometrics between sexes of the two considered species, *C. plumbeus* and *P. glauca*.

Author Contributions: Conceptualization, C.M., methodology, C.M. and F.S.; formal analysis, C.M.; investigation, C.M., F.S., A.N. and U.S.; data curation, C.M. and F.S.; writing—original draft preparation, C.M.; writing—review and editing, F.S., A.N., U.S., A.V., R.T.B. and L.M.; supervision, F.S. and L.M.; project administration, F.S. and L.M. All the authors contributed to writing and finalizing the manuscript. All authors have read and agreed to the published version of the manuscript.

Funding: This research received no external funding.

Institutional Review Board Statement: Not applicable.

Data Availability Statement: Not applicable.

Acknowledgments: We thank all contributors to this research. We would like to thank all our collaborators involved in the MEDLEM project and all the people who provided useful information, in particular, the Italian Coast Guard for helping us to record the information collected during their institutional programs. We thank M.A.L. Zuffi (University of Pisa) for having provided the graphical format of Figure 5 and helped in the statistical analyses. We also thank the direction of Museum of Natural History of Pisa University and Museum of Natural History of Florence University, zoological section, La Specola, for preserving some specimens in the museum collections.

Conflicts of Interest: The authors declare no conflict of interest.

References

1. Compagno, L.J.V. FAO species catalogue. In *Vol. 4. Sharks of the World. An Annotated and Illustrated Catalogue of Sharks Species Known to Date. Part 1. Hexanchiformes to Lamniformes*; FAO Fisheries Synopsis, (125); FAO: Tiburon, CA, USA, 1984; Volume 4, p. 249.
2. Compagno, L.J.V. FAO species catalogue. In *Vol. 4. Sharks of the World. An Annotated and Illustrated Catalogue of Sharks Species Known to Date. Part 2. Carcharhiniformes*; FAO Fisheries Synopsis, (125); FAO: Tiburon, CA, USA, 1984; Volume 4, pp. 251–655.

3. Séret, B.; Serena, F. *The Mediterranean Chondrichthyan Fishes (Sharks, Rays, Skates and Chimaeras): Status and Priorities for Conservation*; Final Report, 25 pp + Annex; UNEP RAC/SP: Tunis, Tunisia, 2002.
4. Serena, F. *Field Identification Guide to the Sharks and Rays of the Mediterranean and Black Sea*; FAO Species Identification Guide for Fishery Purposes; FAO: Rome, Italy, 2005; p. 97, 11 colour plates + egg cases.
5. FAO. *Species Photographic Plates. Mediterranean Sharks*; Barone, M., Serena, F., Dimech, M., Eds.; FAO: Rome, Italy, 2018.
6. FAO. *Species Photographic Plates. Mediterranean Skates, Rays and Chimaeras*; Barone, M., Serena, F., Dimech, M., Eds.; FAO: Rome, Italy, 2018.
7. Serena, F.; Abella, A.J.; Bargnesi, F.; Barone, M.; Colloca, F.; Ferretti, F.; Fiorentino, F.; Jenrette, J.; Moro, S. Species diversity, taxonomy and distribution of Chondrichthyes in the Mediterranean and Black Sea. *Eur. Zool. J.* **2020**, *87*, 497–536. [CrossRef]
8. Stevens, G.; Fernando, D.; Dando, M.; Notarbartolo Di Sciara, G. *Guide to the Manta & Devil Rays of the World*; Wild Nature Press: Plymouth, UK, 2018; p. 144.
9. Carpentieri, P.; Nastasi, A.; Sessa, M.; Srour, A. (Eds.) *Incidental Catch of Vulnerable Species in Mediterranean and Black Sea Fisheries—A Review*; General Fisheries Commission for the Mediterranean. Studies and Reviews. No. 101; FAO: Rome, Italy, 2021; p. 320. [CrossRef]
10. Bonfil, R. Trends and patterns in world and Asian Elasmobranch Fisheries. In *Elasmobranch Biodiversity, Conservation and Management. Proceedings of the International Seminar and Workshop, Sabah, Malaysia, 8–10 July 1997*; Occasional Paper of the IUCN Species Survival Commission; Fowler, S.L., Reed, T.M., Eds.; IUCN: Gland, Switzerland, 2002; pp. 15–32.
11. Mejuto, J.; Garcia-Cortes, B.; de la Serna, J.M. Preliminary scientific estimations of by-catch landed by the Spanish surface longline fleet in 1999 in the Atlantic Ocean and Mediterranean Sea. *ICCAT Collect. Vol. Sci. Pap.* **2002**, *54*, 1150–1163.
12. Peristeraki, P.; Kypraios, N.; Lazarakis, G.; Tserpes, G. By-catches and discards of the Greek swordfish fishery. *ICCAT Collect. Vol. Sci. Pap.* **2008**, *62*, 1070–1073.
13. Damalas, D.; Megalofonou, P. Occurrences of large sharks in the open waters of the southeastern Mediterranean Sea. *J. Nat. Hist.* **2012**, *46*, 2701–2723. [CrossRef]
14. Garibaldi, F. Bycatch in the mesopelagic swordfish longline fishery in the Ligurian Sea (western Mediterranean). *ICCAT Collect. Vol. Sci. Pap.* **2015**, *71*, 1495–1498.
15. Poisson, F.; Arnaud-Haond, S.; Demarq, H.; Cornella, D.; Wendling, B. Satellite tagging of blue shark and pelagic stingray for post release survival and habitat use studies in the Mediterranean Sea. In Proceedings of the Report of the 2016 Intersessional Meeting of the Shark Species Group, Madeira, Portugal, 25–29 April 2016; SCRS/P/2016/020.
16. Mancusi, C.; Serena, F. (Eds.) The Data Base of the MEDiterranean Large Elasmobranchs Monitoring (MEDLEM); Electronic version available and updated on 31 March 2014. Available online: http://www.arpat.toscana.it/temi-ambientali/acqua/acque-marine-e-costiere/medlem (accessed on 31 March 2014).
17. Serena, F.; Barone, M.; Mancusi, C.; Magnelli, G.; Vacchi, M. The MEDLEM database Application: A tool for storing and sharing the large shark's data collected in the Mediterranean countries. In *Proceedings of the Workshop on Mediterranean Cartilaginous Fish with Emphasis on Southern and Eastern Mediterranean*; Basusta, N., Keskin Serena, F., Seret, B., Eds.; Turkish Marine Research Foundation: Istanbul, Turkey, 2006; Volume 23, pp. 118–129.
18. Gallo, S.; Rigers, B.; Bottaro, M.; Cetkovic, I.; Enajjar, S.; Giglio, G.; Lanteri, L.; Mancusi, C.; Barash, A.; Carbonara, P.; et al. Sigthings of large elasmobranchs from the Mediterranean: New data from MEDLEM database in the last five years (2017–2022). In Proceedings of the IEEE International Workshop on Metrology for the Sea; Learning to Measure Sea Health Parameters (MetroSea), Milazzo, Italy, 3–5 October 2022; IEEE: Milazzo, Italy, 2022. [CrossRef]
19. Mancusi, C.; Marsili, L.; Terracciano, G.; Ventrella, S. L'osservatorio Toscano Biodiversità: 2007–2016, dieci anni di attività di recupero cetacei, tartarughe e grandi pesci cartilaginei. In *Proceedings of the Seventh International Symposium: Monitoring of Mediterranean Coastal Areas: Problems and Measurement Techniques, Livorno, Italy, 19–21 June 2018*; Benincasa, F., Ed.; Firenze University Press: Toscana, Italy, 2018; pp. 302–310, ISBN 978-88-6453-811-2.
20. Garibaldi, F.; Orsi Relini, L. Abbondanza estiva, struttura di taglia e nicchia alimentare della verdesca, Prionace glauca, nel Santuario pelagico del Mar Ligure. *Biol. Mar. Mediter.* **2000**, *7*, 324–333.
21. Serena, F.; Barone, M.; Colloca, F.; Vacchi, M. Elasmobranchii, Holocephali. In *Checklist of the Italian Fauna*; Version 1.0; Bologna, M.A., Zapparoli, M., Oliverio, M., Minelli, A., Bonato, L., Cianferoni, F., Stoch, F., Eds. Last Update: 31 May 2021. Available online: https://www.lifewatchitaly.eu/en/initiatives/checklist-fauna-italia-en/checklist/ (accessed on 31 May 2021).
22. Barone, M.; Mazzoldi, C.; Serena, F. *Sharks, Rays and Chimaeras in Mediterranean and Black Seas Key to Identification*; FAO: Rome, Italy, 2022. [CrossRef]
23. Serena, F.; Mancusi, C.; Barone, M. *MEDiterranean Large Elasmobranch Monitoring. Protocollo di Acquisizione Dati*; SharkLife Program: Roma, Italy, 2014; p. 130.
24. Mancusi, C.; Baino, R.T.; Fortuna, C.; De Sola, L.; Morey, G.; Bradai, M.; Kallianotis, A.; Soldo, A.; Hemida, F.; Saad, A.; et al. MEDLEM database, a data collection on large Elasmobranchs in the Mediterranean and Black seas. *Mediter. Mar. Sci.* **2020**, *21*, 276–288. [CrossRef]
25. Last, P.R.; White, W.T.; De Carvalho, M.R.; Séret, B.; Stehmann, M.F.W.; Naylor, G. *Rays of the World*; CSIRO Publishing & Cornell University Press: Ithaca, NY, USA, 2016; p. 790.
26. Ebert, D.A.; Dando, M.; Fowler, S. *Sharks of the World: A Complete Guide*; Princeton University Press: Princeton, NJ, USA, 2021; p. 608. [CrossRef]

27. Eschmeyer, W.N.; Fricke, R.; Van Der Laan, R. (Eds.) *Catalog of Fish: Genera, Species, References*; California Academy of Sciences: San Francisco, CA, USA, 2016. Available online: http://researcharchive.calacademy.org/research/ichthyology/catalog/fishcatmain.asp (accessed on 2 October 2017).
28. Nosal, D.; Cartamil, P.; Wegner, N.C.; Lam, C.H.; Hastings, P.A. Movement ecology of young-of-the-year blue sharks Prionace glauca and shortfin makos Isurus oxyrinchus within a putative binational nursery area. *Mar. Ecol. Prog. Ser.* **2019**, *623*, 99–115. [CrossRef]
29. Compagno, L.J.V.; Dando, M.; Fowler, S. *A Field Guide to the Sharks of the World*; Collins: New York, NY, USA, 2005; p. 368.
30. Carlsson, J.K. Occurrence of neonate and juvenile sandbar sharks, Carcharhinus plumbeus, in the northeastern Gulf of Mexico. *Fish. Bull.* **1999**, *97*, 387–391.
31. Erguden, D.; Kabasakal, H.; Kabakli, F. Young-of-the-year sandbar shark, Carcharhinus plumbeus (Nardo, 1827) (Carcharhiniformes: Carcharhinidae), caught in Iskenderun Bay. *FishTaxa* **2020**, *18*, 18–22.
32. Cailliet, G.M.; Martin, L.K.; Harvey, J.T.; Kusher, D.; Welden, B.A. Preliminary studies on the age and growth of blue, Prionace glauca, com- mon thresher, Alopias vulpinus, and shortfin mako, Isurus oxyrinchus, sharks from California waters. In *Proceedings of the International Work- Shop on Age Determination of Oceanic Pelagic Fishes: Tunas, Billfishes, and Sharks*; Prince, E.D., Pulos, L.M., Eds.; NOAA Tech. Rep. NMFS 8; U.S. Dept. Comm.: Washington, DC, USA, 1983; pp. 179–188.
33. Kwang-Ming, L.; Che-Tsung, C.; Tai-Hsiang, L.; Shoou-Jeng, J. Age, Growth, and Reproduction of the Pelagic Thresher Shark, Alopias pelagicus in the Northwestern Pacific. *Copeia* **1999**, *1*, 68–74.
34. Ebert, D.A. Biological aspects of the sixgill shark, Hexanchus griseus. *Copeia* **1986**, *1*, 131–135. [CrossRef]
35. McFarlane, G.A.; King, J.R.; Saunders, M.W. Preliminary study on the use of neural arches in the age determination of bluntnose sixgill sharks (Hexanchus griseus). *Fish. Bull.* **2002**, *100*, 861–864.
36. Notarbartolo di Sciara, G.; Bianchi, I. *Guida Degli Squali e delle Razze del Mediterraneo*; Franco Muzzio Editore: Roma, Italy, 1998; p. 388.
37. Carrera-Fernández, M.; Galván-Magaña, F.; Ceballos-Vázquez, B.P. Reproductive biology of the blue shark Prionace glauca (Chondrichthyes: Carcharhinidae) off Baja California Sur, México. *Aqua. Int. J. Ichthyol.* **2010**, *16*, 101–110.
38. Druon, J.-N.; Campana, S.; Vandeperre, F.; Hazin, F.H.V.; Bowlby, H.; Coelho, R.; Queiroz, N.; Serena, F.; Abascal, F.; Damalas, D.; et al. Global-Scale Environmental Niche and Habitat of Blue Shark (Prionace glauca) by Size and Sex: A Pivotal Step to Improving Stock Management. *Front. Mar. Sci.* **2022**, *9*, 828412. [CrossRef]
39. Nakano, H. Age, reproduction and migration of blue shark in the North Pacific Ocean. *Bull. Natl. Res. Inst. Far Seas Fish.* **1994**, *31*, 141–256.
40. Pratt, H.L. Reproduction of the blue shark, Prionace glauca. *Fish. Bull.* **1979**, *77*, 445–470.
41. Ferretti, F.; Myers, R.A.; Serena, F.; Lotze, H. Loss of large predatory sharks from the Mediterranean Sea. *Conserv. Biol.* **2008**, *22*, 952–964. [CrossRef]
42. Otero, M.; Serena, F.; Gerovasileiou, V.; Barone, M.; Bo, M.; Arcos, J.M.; Vulcano, A.; Xavier, J. *Identification Guide of Vulnerable Species Incidentally Caught in Mediterranean Fisheries*; IUCN: Malaga, Spain, 2019; p. 204.
43. Öztürk, B. National action plan for the conservation of cartilaginous fishes in the Turkish water of the eastern Mediterranean Sea. *J. Black Sea Medit. Environ.* **2018**, *24*, 91–96.
44. Bradaï, M.N.; Saïdi, B.; Bouaïn, A.; Guélorget, O.; Capapé, C. The Gulf of Gabès (Central Mediterranean): Nursery area for the sandbar shark, Carcharhinus plumbeus (Nardo, 1827) (Chondrichthyes: Carcharhinidae). *Ann. Ser. Hist. Nat.* **2005**, *15*, 187–194.
45. Saïdi, B.; Bradaï, M.N.; Bouaïn, A.; Guélorget, O.; Capapé, C. The reproductive biology of the sandbar shark, Carcharhinus plumbeus (Chondrichthyes: Carcharhinidae), from Gulf of Gabès (southern Tunisia, central Mediterranean). *Acta Adriat.* **2005**, 46+, 47–62.
46. Capapé, C. Nouvelles données sur la morphologie et la biologie de la reproduction de Carcharhinus plumbeus (Nardo, 1827) (Pisces, Carcharhinidae) des côtes Tunisiennes. *Investig. Pesq.* **1984**, *48*, 115–137.
47. Ebert, D.A.; Stehmann, M.F.W. *Sharks, Batoids, and Chimaeras of the North Atlantic*; FAO Species Catalogue for Fishery Purposes, No. 7; FAO: Rome, Italy, 2013; p. 523.
48. Branstetter, S. Early life-history implications of selected carcharhinoid and lamnoid sharks of the northwest Atlantic. In *Elasmobranchs as Living Resources:Advances in Biology, Ecology, Systematics and the Status of the Fisheries*; NOAA Technical Report, 90, Pratt, H.L., Gruber, S.H., Taniuchi, T., Eds.; National Marine Fisheries Service: Silver Spring, MD, USA, 1990; pp. 17–28.
49. Heupel, M.; Carlson, J.K.; Simpfendorfer, C.A. Shark nursery areas: Concepts, definition, characterization and assumptions. *Mar. Ecol. Prog. Ser.* **2007**, *337*, 287–297. [CrossRef]
50. Serena, F.; Roberto Silvestri, R. Preliminary observations on juvenile shark catches as by-catch of the Italian fisheries with particular attention to the Tuscany coasts. In Proceedings of the Codice Armonico, Settimo Congresso di Scienze Naturali Ambiente Toscano, Rosignano, Livorno, Italy, 11 October 2018; pp. 158–169.

diversity

MDPI

Article

Trophic Ecology during the Ontogenetic Development of the Pelagic Thresher Shark *Alopias pelagicus* in Baja California Sur, Mexico

Clara Sánchez-Latorre [1], Felipe Galván-Magaña [1,*], Fernando R. Elorriaga-Verplancken [1], Arturo Tripp-Valdez [1], Rogelio González-Armas [1], Alejandra Piñón-Gimate [1] and Antonio Delgado-Huertas [2]

1 Instituto Politécnico Nacional, Centro Interdisciplinario de Ciencias Marinas, Avenida IPN, s/n Colonia Playa Palo de Santa Rita, La Paz 23096, Baja California Sur, Mexico; clara10sanchez@hotmail.com (C.S.-L.); felorriaga@ipn.mx (F.R.E.-V.); atrippv@ipn.mx (A.T.-V.); rarmas@ipn.mx (R.G.-A.); apinong@ipn.mx (A.P.-G.)

2 Instituto Andaluz de Ciencias de la Tierra, CSIC-UGR, 1808 Granada, Spain; antonio.delgado@csic.es

* Correspondence: galvan.felipe@gmail.com

Abstract: The trophic ecology of the Pelagic Thresher shark (*Alopias pelagicus*) was evaluated based on chemical ecology using stable isotope ratios of carbon ($\delta^{13}C$) and nitrogen ($\delta^{15}N$) in the vertebrae and muscles. Individuals were caught between August 2013 and October 2019 on both the coasts of Baja California Sur, Mexico. In Bahía Tortugas, the mean vertebrae (n = 35) values were 12.72 ± 1.06‰ ($\delta^{15}N$) and −14.79 ± 0.61‰ ($\delta^{13}C$), while in muscles (n = 32) these values were 16.63 ± 0.76‰ ($\delta^{15}N$) and −17.18 ± 0.39‰ ($\delta^{13}C$). In Santa Rosalía, the mean vertebrae (n = 125) isotopic values were 14.4 ± 1.59‰ ($\delta^{15}N$) and −14.18 ± 0.51‰ ($\delta^{13}C$), while in muscles (n = 43), these values were 18.08 ± 0.96‰ ($\delta^{15}N$) and −16.43 ± 0.34‰ ($\delta^{13}C$). These results show higher $\delta^{15}N$ values in Santa Rosalía as an effect of baseline isotopic differences between the two regions, whereas the $\delta^{13}C$ values were lower in Bahía Tortugas, suggesting offshore ecological behavior ($p < 0.05$). In Santa Rosalía, there were significant differences by sex for $\delta^{15}N$ in muscle, whereas the $\delta^{13}C$ showed ontogenetic shifts, indicating that neonates feed in coastal areas more commonly than juveniles or adults ($p < 0.05$). Neither sex nor ontogenetic differences were observed in Bahía Tortugas ($p > 0.05$), suggesting a high overlap between their isotopic niches. Therefore, *Alopias pelagicus* uses the same ecological niche throughout its life, and there is consistency between sexes. The mean trophic position for both tissues and regions was 4.5, which corresponds to a tertiary predator, without any differences between stages or sex. Due to their higher energetic needs, juveniles and females showed the greatest isotopic niche amplitude; thus, their ecological niche is the widest.

Keywords: chemical ecology; stable isotopes; vertebrae; muscle; trophic shifts

1. Introduction

The Pelagic Thresher shark *Alopias pelagicus* (Nakamura inhabits tropical and temperate oceanic waters throughout the Indo-Pacific Ocean, including the eastern coast from Mexico to northern Peru, with no records in the Atlantic Ocean. It is found up to 300 m deep, and its length can reach up to 4.28 m [1]. Thresher sharks are characterized by their long tail, which measures half their body length, and that they use to corner and to disorient and stun the fish and pelagic invertebrates that are part of their diet. This species is considered to be oophagous, due to the egg capsules and fragments of shell found in the stomachs of embryos [2]. Liu et al. [3] established that the Pelagic Thresher shark has two embryos per litter, with a long gestation period of nine months, and they mature at the age of eight years, which suggests that this species is extremely vulnerable to overexploitation and in need of close monitoring.

Mexican fisheries land some of the largest shark catches in the world, dominated mainly by mustelids, but other species such as *A. pelagicus* also represent a significant component [4]. Fisheries along the coasts of Baja California are diverse and opportunistic, while *A. pelagicus* and *Isurus oxyrinchus* (Rafinesque) are specifically targeted for their meat and fins [4].

There is scarce knowledge regarding their biology, with few studies carried out into their age, growth, and reproduction in the northwestern Pacific [3,5] and Ecuador [2]. The trophic ecology of *A. pelagicus* has mainly been examined in the waters off Ecuador [6–8], with Lara et al.'s [9] being the only dietary study in Mexican waters. Polo-Silva et al. [6] found the giant squid *Dosidicus gigas* (d'Orbigny to be the main prey in its diet, followed by the lanternfish *Benthosema panamense* (Tåning) and the squid *Sthenoeuthis oulaniensis* (Lesson). Similar results were obtained by Calle-Morán [7], who reported as the main prey the Red Flying Squid *Ommastrephes bartramii* (Lesueur), the Jumbo Squid *D. gigas*, the Purpleback Fying Squid *S. oualaniensis*, and the South Pacific Hake *Merluccius gayi* (Guichenot). Garcia-Olvera [10] reported anchovy *Engraulis mordax* (Girard) to be its main prey, followed by the family of lanternfishes Myctophidae, and *Paralabrax* spp. (Girard). The same topic has been recently studied in Indonesia for the first time [11], with the following identified as the three top prey species: the Frigate Tuna *Auxis thazard* (Lacepède), the Purpleback Flying Aquid *S. oualaniensis*, and the Spiny Lanternfish *Dasyscopelus spinosus* (Steindachner).

The most common method used to analyze prey is to extract the stomach and identify stomach contents to the lowest taxonomical level possible, but it is also possible to obtain samples via non-lethal methods [12]. One technique that has been used previously is stomach flushing, also referred to as gastric lavage. This technique involves pumping water via a tube down the throat of the animal into the stomach, expelling the stomach contents via the mouth [12]. Another less widely used method is applying forceps to evert the stomachs of small sharks. However, since this involves either restraining the shark upside down, or anesthetizing the animal before reaching into its mouth, it is not appropriate for larger species [12].

Stable isotope analysis, particularly using nitrogen and carbon, allows us to evaluate the structure and dynamics of ecological communities, with this approach combining the benefits of both trophic-level and food web paradigms in food web ecology [13]. Moreover, this method is less invasive than others previously mentioned, since stable isotope samples can be taken by extracting blood or by performing biopsies [14]. Values of $\delta^{15}N$ are used to estimate trophic levels and breadth; consumers are isotopically enriched by 3–4‰ relative to their prey [13,15]. In contrast, values of $\delta^{13}C$ remain relatively unaffected by trophic level (0.5–1‰), providing information about trophic habitat use across the inshore-offshore gradient [15,16].

For slow-growing species such as sharks, the isotopic signals given off by the muscle tissue provide information about the prey assimilated by the predator 1–2 years prior to its consumption [14,16]. Other tissues such as vertebrae are metabolically inert and contain growth layers that can record ecological information over the lifespan of a single individual [16,17]. Ontogenetic shifts in the use of trophic habitat often reflect changes in survival strategies. While juveniles use their energy to grow, adults prioritize activities such as reproduction [18]. These changes in feeding habits are common in sharks and rays, as these species are able to shift their prey types and feeding areas as they grow [19]. Understanding the ecological life history of pelagic species is important to ensure their survival, especially in long-living species such as elasmobranchs [18].

The combined assessment of trophic ontogeny in *A. pelagicus* from multiple regions, focusing on different tissues and life stages, allows us not only to analyze their feeding behaviors, but also to isotopically compare tissues with different metabolic rates and relate these to areas with unique oceanographic conditions.

Considering that stable isotopes of the Pelagic Thresher shark have only been only studied in Ecuadorian waters, the aim of this study is to examine the trophic ontogeny of this species in both the coastal areas of Baja California Sur, Mexico, by analyzing the nitrogen and carbon stable isotopes in vertebrae and muscles.

2. Materials and Methods

2.1. Study Area and Sample Collection

Samples were collected in two areas: the fishing town of Bahía Tortugas (27°39′35″ N; 114°52′35″ W), located on the western coast of the Baja California Sur Peninsula, and Santa Rosalía (27°20′20″ N; 112°16′01″ W), located on the eastern coast of Baja California Sur (Figure 1).

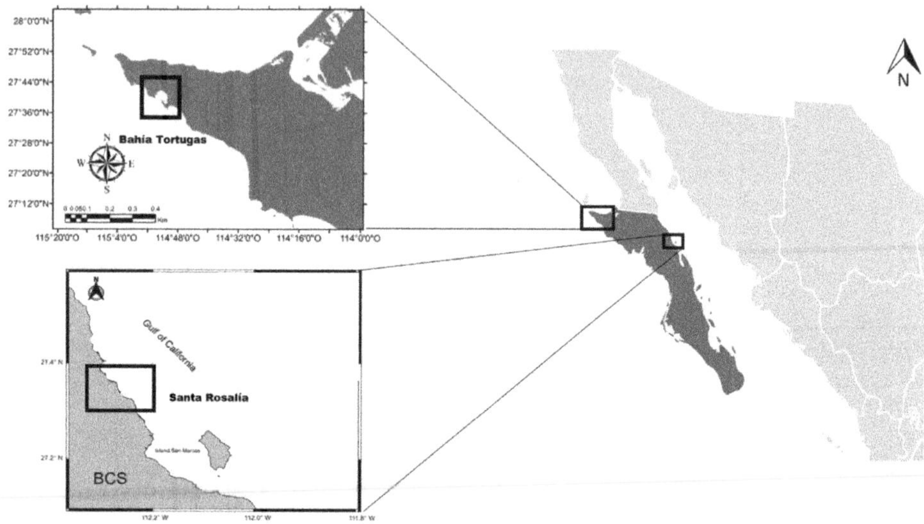

Figure 1. Location of two study areas, Bahia Tortugas and Santa Rosalía in Baja California Sur, Mexico.

Bahía Tortugas (BT) is known for the frequent upwelling phenomenon that occurs there, which provides nutrients to the surface that remain available for primary production. Therefore, this region exhibits a high phyto- and zooplankton biomass, which results in the high productivity of fishery resources [20].

Santa Rosalia (SR) is located in the Gulf of California, the only evaporation basin of the Pacific Ocean, due to its location between two hot land masses and the absence of freshwater inflow to the region. The variable depth of the Gulf of California, the characteristics of its habitat, and its unique location in a transition zone between temperate and tropical faunal regions endows the area with a unique biological richness [21].

In BT, samples of *A. pelagicus* were collected using artisanal long-line fishery equipment from August 2013 to August 2016. In SR, samples were collected by fishermen using gill nets from October 2017 to October 2019. The total and precaudal length (TL and PL) were measured and the sex was determined by the presence of claspers in males. Sexual maturity in males was established via the size and condition of the clasper (rotation, calcification, and semen presence) and the development of the testes. Males were divided into two reproductive stages. Juvenile/Immature: short and non-calcified clasper; testes soft, elongated, and not lobated. Adult/Mature: calcified claspers with fully lobated testes.

Approximately 20 g of muscle along with one or more vertebrae from the dorsal region near the head was sampled from each individual organism. The samples were stored in polyethylene bags properly identified and transported on ice to the CICIMAR-IPN. They were frozen in the laboratory until analysis.

2.2. Laboratory Analysis

The vertebrae were defrosted, cleaned, and dried. The radius of each vertebra was measured using a digital vernier and related to the precaudal length by linear regression. Based on this equation and on the size at maturity, as proposed by Romero-Caicedo et al. [2], three ontogenetic stages in each vertebra were determined. For neonates, the first visible growth layer was sampled. Juvenile samples were those taken from the far end of the growth layer—up to 8.6 mm radius for males and 8.7 mm radius for females. For mature sharks, samples were taken from the outer edge of the vertebra when the radii exceeded 8.6 and 8.7 mm, respectively. Thus, for each vertebra of a mature shark, three samples were taken, while for each vertebra of an immature shark, two samples were taken. The samples were extracted using a microdrill with a 1 mm bit and were exposed for 24 h to a hydrochloric acid steam bath to remove inorganic carbon.

Approximately 5 g of each muscle sample was put into vials. As elasmobranchs retain urea and fat in their tissues, which can influence $\delta^{15}N$ and $\delta^{13}C$ values (depleting their heavy isotope values), these products had to be removed. Urea was extracted following Kim and Koch's [22] methods, using a mechanic bath (Bransonic M 8800) wherein each sample was washed three times with 10 mL deionized water for 15 min. The samples were then lyophilized at 0.123 mbar and −40 °C for 48 h, and then ground and homogenized in an agate mortar. Lipids were not removed since Post et al. [23] suggested an arithmetic correction that could be used to remove lipids from the isotopic signature, as follows:

$$\delta^{13}C_{\ corrected} = \delta^{13}C_{\ sample} - 3.32 + 0.99 \times C{:}N \tag{1}$$

The $\delta^{13}C$ and $\delta^{15}N$ values were determined at the Instituto Andaluz de Ciencias de la Tierra in Granada, Spain, using a DELTA plus XL, Thermo-Finnagen isotope ratio mass spectrometer (IRMS, Bremen, Germany). The isotopic results are expressed as δ values:

$$\delta^{13}C \text{ or } \delta^{15}N = 1000 \times [(R_{sample}/R_{standard}) - 1] \tag{2}$$

where R_{sample} and $R_{standard}$ are the $^{13}C/^{12}C$ or $^{15}N/^{14}N$ ratios of the sample and standard, respectively. The standards were Vienna-Pee Dee Belemnite limestone for carbon and atmospheric N_2 for nitrogen. The units are expressed as parts per thousand (per mil, ‰).

2.3. Data Analysis

The normality of data was assessed by the Kolmogorov–Smirnov–Lilliefors test and variance homogeneity by the Levene test. These were used to test the null hypothesis that a set of data originated from a normal and homoscedastic distribution. If they did, we used a parametrical test to assess for significant differences in mean $\delta^{13}C$ and $\delta^{15}N$ between sexes, stages, and regions. In this case, the most appropriate test was a three-factor analysis of variance (ANOVA), considering that we had three independent categorical variables and one dependent continuous variable. This test reveals whether the variance arose by chance or was the influence by the factors. If the data were not normally distributed, we used a Kruskal–Wallis test, which is accurate when applied to non-parametric data with three categorical variables. Moreover, using this non-parametric test allowed us to perform statistical analysis without necessarily transforming the data.

Estimates of the trophic position (TP) were calculated using the R package tRophicPosition [24]. A different base organism was used for each region: particulate organic matter (POM) for SR ($\delta^{15}N_{POM}$= 11.1‰, TP_{POM} = 1) [25] and the pelagic red crab *Pleuroncodes planipes* (Stimpson) ($\delta^{15}N_{Pleuroncodes\ planipes}$ = 12.10‰, $TP_{Pleuroncodes\ planipes}$ = 2) [20] for BT. As the trophic discrimination factor (TDF), 1.95‰ was used for vertebral tissue and 2.44‰ for muscle tissue [14].

To determine the niche breadth and trophic overlap between stages, sexes, and regions, we used the package SIBER (Stable Isotope Bayesian Ellipses in R) from the program R [26]. This analysis uses measurements based on ellipses calculated by a covariance matrix that defines their area (Standard Ellipse Corrected Area, SEAc) to show the trophic niche breadth. Using this method, it is possible to obtain the overlap between ellipses, whereby values close to 1 represent high trophic overlap [26].

3. Results

For BT, a total of 35 vertebrae and 32 muscle samples were used. For SR, a total of 125 vertebrae and 32 muscle samples were used.

Non-parametrical tests were applied to the $\delta^{13}C$ vertebrae data in BT (F = 8.405, $p = 0.007$), as well as to the $\delta^{15}N$ (D = 0.094, $p = 0.007$) and $\delta^{13}C$ (F = 6.178, $p = 0.003$) vertebrae data in SR, since these did not follow a normal or homoscedastic distribution.

The $\delta^{15}N$ values in vertebrae and muscles were higher in SR than in BT ($X^2 = 37.848$, $p = 7.648 \times 10^{-10}$ and F = 49.54, $p = 8.84 \times 10^{-10}$, respectively), and the $\delta^{13}C$ also presented less negative vertebrae and muscle values in SR ($X^2 = 26.854$, $p = 2.194 \times 10^{-7}$ and F = 75.8, $p = 6.62 \times 10^{-13}$, respectively) (Figure 2; Table 1). In BT and SR, the isotopic niches presented an overlap of 39% in the vertebrae and an overlap of 23% in the muscles. Moreover, the $\delta^{15}N$ values were 4‰ higher in the muscles than in the vertebrae ($X^2 = 125.07$, $p < 2.2 \times 10^{-16}$) (Figure 2; Table 1).

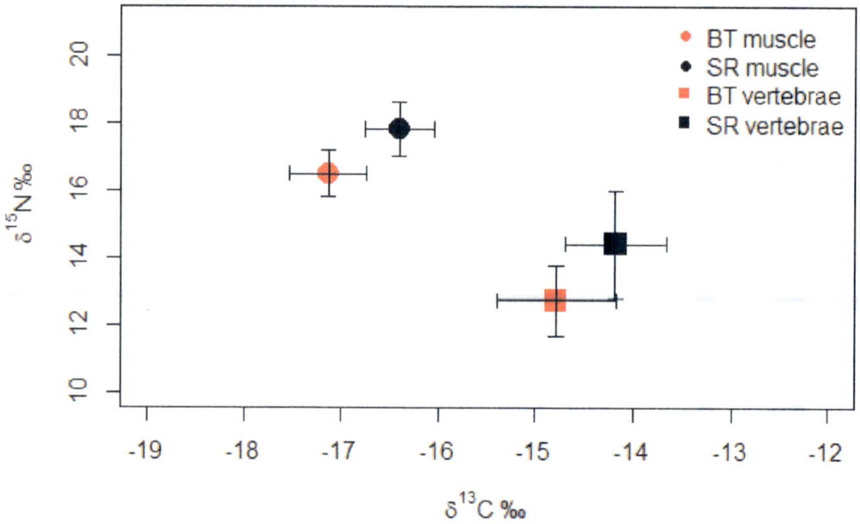

Figure 2. $\delta^{15}N$ and $\delta^{13}C$ values (mean ± SD ‰) of *A. pelagicus* in Bahía Tortugas (BT) and Santa Rosalía (SR) in vertebrae and muscles.

Table 1. Summary of $\delta^{15}N$ and $\delta^{13}C$ values (mean ± SD ‰) by maturity stage and sex in vertebrae and muscles of *A. pelagicus* in Bahía Tortugas (BT) and Santa Rosalía (SR).

	Vertebrae BT (n = 35)		**Muscle BT (n = 32)**		**Vertebrae SR (n = 125)**		**Muscle SR (n = 43)**	
	$\delta^{15}N$	$\delta^{13}C$	$\delta^{15}N$	$\delta^{13}C$	$\delta^{15}N$	$\delta^{13}C$	$\delta^{15}N$	$\delta^{13}C$
Neonates	12.67 ± 1.17	−14.6 ± 0.53			14.91 ± 1.59	−13.84 ± 0.41		
Juveniles	12.74 ± 1.18	−14.97 ± 0.68	16.83 ± 0.79	−17.26 ± 0.46	14.47 ± 0.87	−14.6 ± 0.42	17.78 ± 0.76	−16.48 ± 0.29
Adults	12.79 ± 0.67	−14.78 ± 0.59	16.51 ± 0.74	−17.12 ± 0.35	14.89 ± 1.15	−14.27 ± 0.51	17.83 ± 0.85	−16.32 ± 0.4
Females	12.92 ± 1.21	−14.89 ± 0.42	16.61 ± 0.81	−17.19 ± 0.39	14.31 ± 1.64	−14.21 ± 0.52	17.95 ± 0.77	−16.42 ± 0.37
Males	12.34 ± 0.56	−14.59 ± 0.85	16.66 ± 0.70	−17.16 ± 0.43	14.67 ± 1.45	−14.08 ± 0.49	17.44 ± 0.74	−16.29 ± 0.33
Mean	12.72 ± 1.06	−14.79 ± 0.61	16.63 ± 0.76	−17.18 ± 0.39	14.4 ± 1.59	−14.18 ± 0.51	18.08 ± 0.96	−16.43 ± 0.34

Statistical tests confirmed the similarity of trophic habits of *A. pelagicus* in BT between maturity stages (neonates, juveniles, and adults) and sexes (females and males) for both tissues (Table 2). In SR, the $\delta^{15}N$ remained constant between stages for both tissues, while the values for female were higher than those for males in the muscles. The $\delta^{13}C$ showed significant differences between stages in vertebral tissue (Table 2). Even when statistical differences were only significant in SR, the $\delta^{13}C$ values were more negative in both regions in juveniles and more positive in neonates (Table 1). However, females presented lower $\delta^{13}C$ values than males in both areas and tissues (Table 1).

The SIBER analysis showed an overlap of between 41% and 68% in all groups (except sexes) for vertebrae in BT (27%) (Figures 3 and 4; Table 3). Juveniles and females presented wider ellipse areas for both tissues in BT and for vertebrae in SR (Table 3).

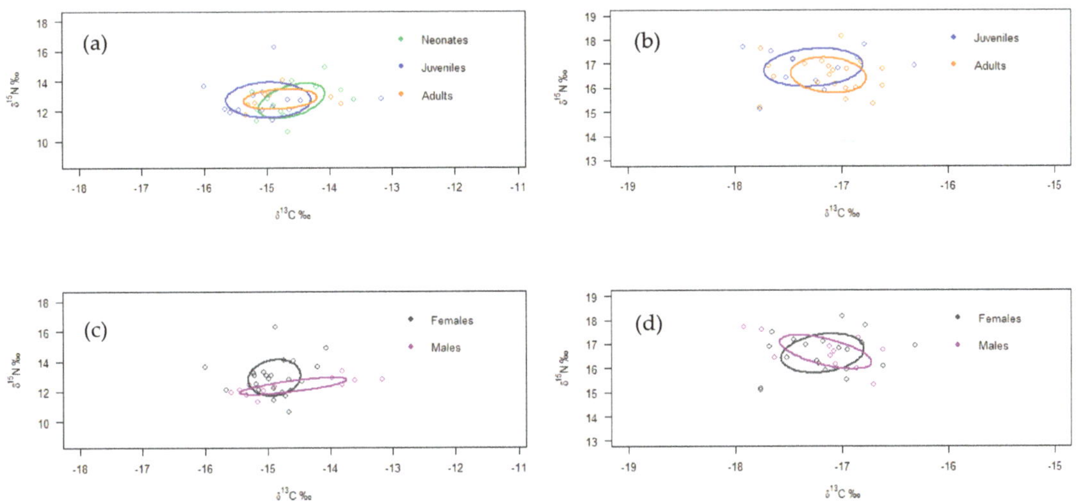

Figure 3. Isotopic niche (ellipses) by stage, sex, and tissue of *A. pelagicus* in Bahía Tortugas (BT). (**a**) Vertebral tissue by stage; (**b**) muscle tissue by stage; (**c**) vertebral tissue by sex (**d**); muscle tissue by sex.

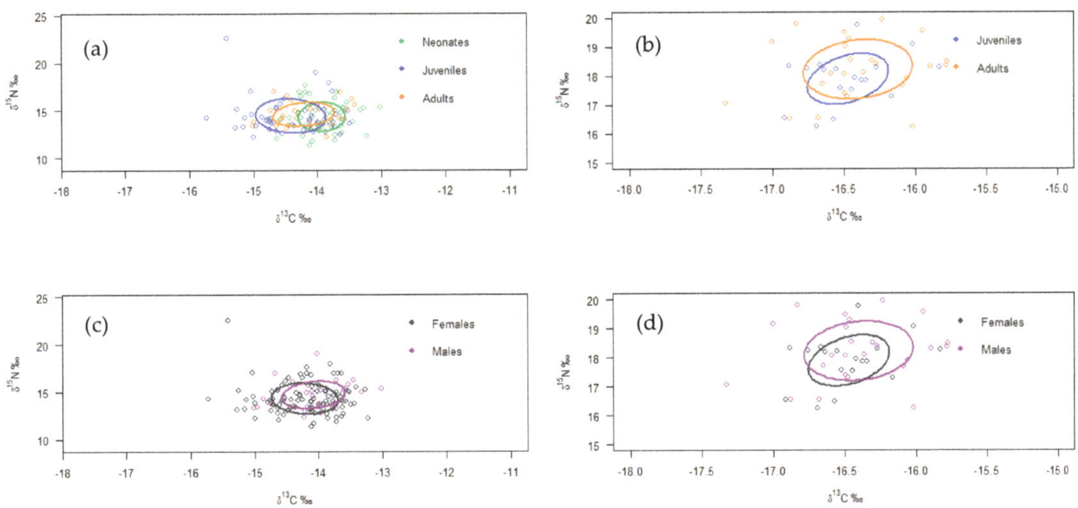

Figure 4. Isotopic niche (ellipses) by stage, sex, and tissue from *A. pelagicus* in Santa Rosalía (SR). (**a**) Vertebral tissue by stage; (**b**) muscle tissue by stage; (**c**) vertebral tissue by sex; (**d**) muscle tissue by sex.

Table 2. $\delta^{15}N$ and $\delta^{13}C$ ANOVA and Kruskal–Wallis analysis of vertebrae and muscles of *A. pelagicus* based on maturity stage and sex in Bahía Tortugas (BT) and Santa Rosalía (SR).

	Vertebra BT				Muscle BT				Vertebra SR				Muscle SR			
	$\delta^{15}N$		$\delta^{13}C$		$\delta^{15}N$		$\delta^{13}C$		$\delta^{15}N$		$\delta^{13}C$		$\delta^{15}N$		$\delta^{13}C$	
	F	p	X^2	p	F	p	F	p	X^2	p	X^2	p	F	p	F	p
Stage	0.22	0.64	2.97	0.23	1.59	0.22	1.19	0.28	1.08	0.58	19.42	6.08×10^{-5}	2.93	0.09	0.38	0.54
Sex	2.37	0.13	0.09	0.75	0.02	0.88	0.007	0.93	2.21	0.14	1.11	0.30	5.68	0.02	0.042	0.84

Table 3. Stable Isotope Bayesian Ellipses in R (SIBER) analysis: Standard Ellipse Corrected Area, SEAc (‰^{2}) and overlap (%) from vertebrae and muscles of *A. pelagicus* based on maturity stages and sex in Bahía Tortugas (BT) and Santa Rosalía (SR).

	Vertebra BT		Muscle BT		Vertebra SR		Muscle SR	
	SEAc	Overlap	SEAc	Overlap	SEAc	Overlap	SEAc	Overlap
Neonates	1.9	48			1.79	49		
Juveniles	2.74	41	1.24	47	3.17	54	0.78	48
Adults	1.4		0.86		1.94		1.27	
Males	1.1	27	0.83	44	2.28	68	1.31	46
Females	1.67		1		2.68		0.9	
	Overlap of neonates and adults: 41%				Overlap of neonates and adults: 55%			

The estimated TP values obtained in BT were 4.6 (vertebrae) and 4.4 (muscle). These values for SR were 4.3 (vertebrae) and 4.5 (muscle). No statistical differences were shown between regions, tissues, sexes, or maturity groups ($p < 0.05$).

4. Discussion

4.1. Comparison between Regions

The Gulf of California presents relatively high baseline $\delta^{15}N$ values due to the denitrification processes that occur in the minimum oxygen zone [27], where ^{14}N is mostly consumed, leaving a ^{15}N-enriched nitrate pool [28]. Thus, $\delta^{15}N$ values of particulate organic matter in the Gulf of California are higher (11.1‰) than those on the western coast of Baja California Sur (8.5‰) [27]. The same pattern was shown for *A. pelagicus* in this study, and so the isotopic contrast between both coasts of Baja California Sur is thought to be related to this difference in the trophic baseline, and not to the fact that the sharks showed different trophic positions on each coast.

Fewer negative $\delta^{13}C$ values are found in productive inshore waters, such as in upwelling regions, while more negative values are found in less productive offshore waters [28]. The results suggest that in BT, *A. pelagicus* feed in offshore habitats, while in SR, this shark species presents a trophic inshore habitat. Another explanation for these differences in $\delta^{13}C$ values is the narrow continental shelf found in BT, which has a strong offshore influence. Hence, even when sharks fed in both regions at similar distances from the coast, those caught on the oceanic shelf (BT) showed more negative $\delta^{13}C$ values. Moreover, environmental conditions such as temperature may also affect the availability of prey in each region. While in summer, the water in SR reaches 30 °C [29], in BT, it stays at 19 °C [30] Thus, the prey in BT come from deeper and colder waters, as compared to the shallower and warmer waters in SR.

Isotopic niches represent the ecological niche based on a determined area inside of the δ space, where the coordinates are $\delta^{13}C$ (environmental components) and $\delta^{15}N$ (trophic components) [31]. The minimal overlap of isotopic niches from both study areas also indicates the different ecological niches of the Pelagic Thresher shark in each region, as discussed above.

4.2. Comparison between Tissues

For both areas, the $\delta^{15}N$ values were higher in muscles than in vertebrae. Likewise, $\delta^{13}C$ values were more positive in vertebrae than in muscles. This isotopic variability among tissues has been previously explained by the fact that each tissue has a different metabolic turnover rate and a different TDF. For example, MacNeil et al. [32] demonstrated that vertebrae presented the slowest $\delta^{15}N$ turnover rate due to slow layer growth. These authors also found that $\delta^{13}C$ decreases with increasing metabolic tissue rates. It has also been observed that muscle presents a higher degree of enrichment (2.44‰) than vertebrae (1.95‰), which could have an influence on the nitrogen isotope ratio [14]. Due to this turnover rate, the vertebrae emit isotopic signals over the lifetime of the organism, while the muscles only yield isotopic information from the last month of the animal's life [17,32].

4.3. Isotopic Analysis by Maturity Stages

No ontogenetic differences in $\delta^{15}N$ values were found in *A. pelagicus*, which indicates that this species feeds on the same prey groups throughout its life. In BT, the $\delta^{13}C$ values also stayed stable throughout the organism's life. Similar results were obtained in Ecuador by Calle-Morán [7], with no ontogenetic differences for any isotope ratio. Lara et al. [9] analyzed the stomach contents of *A. pelagicus* in BT and determined that the sardine *Sardinops sagax* (Jenyns was the principal prey for mature and immatures sharks, suggesting the same feeding pattern despite maturity stage, thus supporting our results. On the other hand, Estupiñán-Montaño [8] showed that the $\delta^{13}C$ values in neonates presented differences from those in juveniles and adults off the Galápagos Islands. Similar results were found in the present study for SR, where the vertebrae presented differences among all maturity stages, from which we can infer a possible ontogenetic inshore–offshore movement for *A. pelagicus*.

Lowe et al. [19] pointed out that sharks change their diet and ecological needs as they grow, thus exploiting different areas, since they are segregated by size. In this study, the findings for BT do not reflect this behavior, while in SR they do, corroborating that *A. pelagicus* shows flexibility in its trophic behavior depending on its habitat and the inherent requirements related to its development.

In both areas, juveniles present the lowest $\delta^{13}C$ values and the widest isotopic niches, while neonates show the highest $\delta^{13}C$ values and most narrow isotopic niches. It can be inferred that on both coasts of Baja California Sur, juvenile individuals of *A. pelagicus* present a wider and more offshore feeding habitat, while neonates present a narrow and more inshore feeding habitat. Calle-Morán [7] attributed this phenomenon to the limitation of neonates reaching offshore waters, causing their habitat to be restricted to inshore waters. Moreover, as juveniles are not completely developed, they might focus most of their energy on growing, and as such, they need more nutrients and probably a wider ecological niche.

4.4. Isotopic Analysis by Sex

No differences in the $\delta^{15}N$ values of *A. pelagicus* by sex were found in BT for any tissue, which is consistent with Lara et al.'s [9] findings in the same area, where males and females presented similar stomach contents. On the other hand, the $\delta^{15}N$ values in SR assessed by sex, were similar in the vertebrae; however, in muscles, they showed significant differences. Once again, a contrast between the tissues is highlighted. These data suggest that male and female *A. pelagicus* in SR were feeding on prey that belonged to the same trophic level during its lifetime. Therefore, the overlap in their isotopic niches is high (68%). Nevertheless, assessments of the muscle indicate that, during the lasts months before their capture, males and females consumed different prey.

The $\delta^{13}C$ values were similar between sexes in both study areas and tissues; however, females presented the lowest values. Moreover, they showed a larger ellipse area than males. Therefore, even if *A. pelagicus* did not segregate by sex on the coasts of Baja California Sur, females presented inclination towards a wider and more oceanic trophic habitat than males. This may indicate that females have greater energy and nutritional needs than males, related to the requirements of pregnancy [16].

It is important to consider that ontogenetic and sex differences were only seen in SR, while in BT, all sharks presented the same trophic behavior, regardless of their sex and size. These results indicate that the ecological habitat in SR is more heterogeneous than in BT, as *A. pelagicus* presented a wider isotopic niche (2.6‰^{2}) during its lifetime in the Gulf of California than on the western coast of Baja California Sur (2.1‰^{2}).

4.5. Trophic Position

Polo-Silva et al. [16] reported that the most appropriate TDF is that published by Kim et al. [33], who carried out a study on the feeding habits of the Leopard Shark *Triakis semifasciata* over 1250 days. Nevertheless, in the present study, we used the TDF proposed by Hussey et al. [14]. Even though the experiment was shorter (around 365 days), these authors developed the factor for vertebrae as well as for muscle.

In both study areas, the TP for the Pelagic Thresher shark was around 4.5, classifying it as a tertiary predator, with no changes by sex or stage. These results agree with those of Lara et al. [9] and Fernández-Aguirre (in process), who reported the sardine *Sardinops sagax* as the main prey for *A. pelagicus* in BT, and anchovy *Engraulis mordax* in SR. Since both these prey feed on plankton, they have a similar TP and consequently, *A. pelagicus* presented a similar TP in both areas.

No variability in the TP by sex or stage was found in this study, which was also reported by Polo-Silva et al. [6], who derived a TP of 3.9 for this species. On the other hand, Calle-Morán [7] obtained the same results (4.5) as we did in our study. Therefore, once more, it can be inferred that *A. pelagicus* shows flexible trophic behavior depending on the availability of prey in the surrounding environment.

5. Conclusions

The present study is the first to analyze the chemical ecology of *A. pelagicus* using stable isotopes in Mexican waters. It provides information on the trophic ontogeny of this species on the western and eastern coasts of Baja California Sur. On the western coast, both isotopes remained stable by sex and stage; thus, this area presented stable feeding habitats for *A. pelagicus*. On the eastern coast, the $\delta^{13}C$ values showed no differences between sexes, and did so only between stages, as neonates probably fed close to the coast. On both coasts, neonates presented a narrow and onshore feeding habit, while juveniles presented a wide and offshore one. *A. pelagicus* was classified as a tertiary predator in our study area.

The use of stable isotopes to measure ontogenetic changes is a very useful tool, especially in environments that are highly variable, such as SR. The use of vertebrae as an indicator of ontogenetic changes is more utile; however, this type of tool is complementary rather than exclusive. For example, the isotopic signatures in muscles show variations that could be explained by the isotopic signatures in the vertebrae. This study makes a novel contribution to the ecological knowledge base (including habitat use) regarding the Thresher Shark in waters off the Mexican Pacific and Gulf of California coasts, providing potential tools for the better management of the species in both areas.

Author Contributions: Conceptualization, C.S.-L., F.G.-M. and F.R.E.-V.; data curation, F.G.-M.; formal analysis, C.S.-L.; investigation, C.S.-L., F.G.-M., F.R.E.-V., A.T.-V., R.G.-A., A.P.-G. and A.D.-H.; methodology, C.S.-L.; resources, F.R.E.-V., A.P.-G. and A.D.-H.; supervision, A.T.-V., R.G.-A., A.P.-G. and A.D.-H.; writing—original draft, C.S.-L.; writing—review and editing, C.S.-L., F.G.-M., A.T.-V., R.G.-A., A.P.-G. and A.D.-H. All authors have read and agreed to the published version of the manuscript.

Funding: Thanks to Instituto Politécnico Nacional, Project SIP-IPN 20220579 for support.

Institutional Review Board Statement: Not applicable.

Data Availability Statement: To get data supporting results contact Clara Sánchez Latorre.

Acknowledgments: R.G.-A., F.G.-M., F.R.E.-V., A.P.-G., and A.T.-V. Thanks to Instituto Politécnico Nacional for fellowships granted (COFAA, EDI).

Conflicts of Interest: The authors declare no conflict of interest.

References

1. Weigmann, S. Annotated checklist of the living sharks, batoids and chimaeras (Chondrichthyes) of the world, with a focus on biogeographical diversity. *J. Fish Biol.* **2016**, *88*, 837–1037. [CrossRef] [PubMed]
2. Romero-Caicedo, A.F.; Galván-Magaña, F.; Martinez-Ortiz, J. Reproduction of the pelagic thresher shark *Alopias pelagicus* in the equatorial Pacific. *J. Mar. Biol. Assoc.* **2014**, *94*, 1501–1507. [CrossRef]
3. Liu, K.-M.; Chen, C.-T.; Liao, T.-H.; Joung, S.-J. Age, Growth, and Reproduction of the Pelagic Thresher Shark, *Alopias pelagicus* in the Northwestern Pacific. *Copeia* **1999**, *1999*, 68–74. [CrossRef]
4. Smith, W.D.; Bizzarro, J.J.; Cailliet, G.M. La pesca artesanal de elasmobranquios en la costa oriental de Baja California, México: Características y consideraciones de manejo. *Cienc. Mar.* **2009**, *35*, 209–236. [CrossRef]
5. Drew, M.; White, W.T.; Dharmadi; Harry, A.V.; Huveneers, C. Age, growth and maturity of the pelagic thresher *Alopias pelagicus* and the scalloped hammerhead Sphyrna lewini. *J. Fish Biol.* **2015**, *86*, 333–354. [CrossRef] [PubMed]
6. Polo-Silva, C.; Rendón, L.; Galván-Magaña, F. Descripción de la dieta de los tiburones zorro (*Alopias pelagicus*) y (*Alopias superciliosus*) durante la época lluviosa en aguas ecuatorianas. *PANAMJAS* **2009**, *4*, 556–571.
7. Calle-Morán, M.D. Ecología Trófica del Tiburón Zorro Pelágico *Alopias pelagicus* en Santa Rosa de Salinas, Pacífico Ecuatoriano. Master's Thesis, Universidad Nacional Autónoma de México, Mexico City, Mexico, 2010.
8. Estupiñán-Montaño, C. Ontogenia Alimentaria de Tres Especies de Tiburones Pelágicos: *Alopias pelagicus*, *Carcharhinus falciformis* y *Prionace glauca* en la Reserva Marina de Galápagos, Ecuador. Master's Thesis, Instituto Politécnico Nacional, Centro Interdisciplinario de Ciencias Marinas, La Paz, Mexico, 2016.
9. Lara, A.; Galván-Magaña, F.; Elorriaga-Verplancken, F.; Marmolejo-Rodríguez, A.J.; González-Armas, R.; Arreola-Mendoza, L.; Sujitha, S.B.; Jonathan, M.P. Bioaccumulation and trophic transfer of potentially toxic elements in the pelagic thresher shark *Alopias pelagicus* in Baja California Sur, Mexico. *Mar. Pollut. Bull.* **2020**, *156*, 111192. [CrossRef]
10. García-Olvera, M.E. Ecología Trófica de los Tiburones Zorro (*Alopias pelagicus*) y Piloto (*Carcharhinus falciformis*) en los Puertos de Santa Rosalía y Punta Lobos, Baja California Sur. Bachelor's Thesis, Universidad Autónoma de Ciudad Juarez, Ciudad Juarez, Mexico, 2023; 41p.
11. Alghozali, F.A.; Salsabila, R.; Gustianto, M.W.D.; Putri, H.M.I.H.; Himawan, M.R.; Yuneni, R.R.; Hatmoro, C.K.; Rezkiani, M. Diet analyses of the pelagic thresher shark, *Alopias pelagicus* (Lamniformes: Alopiidae), from the Lombok Strait waters, Indonesia. *Biodiversitas J. Biol. Divers.* **2023**, *24*, 3708–3714. [CrossRef]
12. Barnett, A.; Redd, K.S.; Frusher, S.D.; Stevens, J.D.; Semmens, J.M. Non-lethal method to obtain stomach samples from a large marine predator and the use of DNA analysis to improve dietary information. *J. Exp. Mar. Biol. Ecol.* **2010**, *393*, 188–192. [CrossRef]
13. Post, D.M. Using stable isotopes to estimate trophic position: Models, methods, and assumptions. *Ecology* **2002**, *83*, 703–718. [CrossRef]
14. Hussey, N.E.; Brush, J.; McCarthy, I.D.; Fisk, A.T. Δ15N and δ13C diet–tissue discrimination factors for large sharks under semi-controlled conditions. *Comp. Biochem. Physiol. Part A Mol. Integr. Physiol.* **2010**, *155*, 445–453. [CrossRef] [PubMed]
15. Peterson, B.J.; Fry, B. Stable Isotopes in Ecosystem Studies. *Annu. Rev. Ecol. Syst.* **1987**, *18*, 293–320. [CrossRef]
16. Polo-Silva, C.; Newsome, S.D.; Galván-Magaña, F.; Grijalba-Bendeck, M.; Sanjuan-Muñoz, A. Trophic shift in the diet of the pelagic thresher shark based on stomach contents and stable isotope analyses. *Mar. Biol. Res.* **2013**, *9*, 958–971. [CrossRef]
17. Estrada, J.A.; Rice, A.N.; Natanson, L.J.; Skomal, G.B. Use of Isotopic Analysis of Vertebrae in Reconstructing Ontogenetic Feeding Ecology in White Sharks. *Ecology* **2006**, *87*, 829–834. [CrossRef] [PubMed]
18. Grubbs, R.D. Ontogenetic shifts in movements and habitat use. In *Sharks and Their Relatives II*; CRC Press: Boca Raton, FL, USA, 2010; pp. 335–366.
19. Lowe, C.G.; Wetherbee, B.M.; Crow, G.L.; Tester, A.L. Ontogenetic dietary shifts and feeding behavior of the tiger shark, *Galeocerdo cuvier*, in Hawaiian waters. *Environ. Biol. Fishes* **1996**, *47*, 203–211. [CrossRef]
20. Martínez-Ayala, J.C. Ecología Trófica del Cazón Mamón *Mustelus lunulatus* en la Costa Occidental de Baja California Sur, México. Master's Thesis, Instituto Politécnico Nacional, Centro Interdisciplinario de Ciencias Marinas, La Paz Mexico, 2018.
21. Bizzarro, J.J.; Smith, W.D.; Hueter, R.E.; Tyminski, J.; Márquez, J.F.; Castillo, J.L.; Cailliet, G.M. *El Estado Actual de los Tiburones y Rayas Sujetos a Explotación Comercial en el Golfo de California: Una Investigación Aplicada al Mejoramiento de su Manejo Pesquero y Conservación*; Moss Landing Marine Laboratories Technical Publication: Moss Landing, CA, USA, 2007.
22. Kim, S.L.; Koch, P.L. Methods to collect, preserve, and prepare elasmobranch tissues for stable isotope analysis. *Environ. Biol. Fishes* **2012**, *95*, 53–63. [CrossRef]
23. Post, D.M.; Layman, C.A.; Arrington, D.A.; Takimoto, G.; Quattrochi, J.; Montaña, C.G. Getting to the fat of the matter: Models, methods and assumptions for dealing with lipids in stable isotope analyses. *Oecologia* **2007**, *152*, 179–189. [CrossRef]
24. Quezada-Romegialli, C.; Jackson, A.L.; Hayden, B.; Kahilainen, K.K.; Lopes, C.; Harrod, C. TRophicPosition, an r package for the Bayesian estimation of trophic position from consumer stable isotope ratios. *Methods Ecol. Evol.* **2018**, *9*, 1592–1599. [CrossRef]

25. Aurioles-Gamboa, D.; Rodríguez-Pérez, M.Y.; Sánchez-Velasco, L.; Lavín, M.F. Habitat, trophic level, and residence of marine mammals in the Gulf of California assessed by stable isotope analysis. *Mar. Ecol. Prog. Ser.* **2013**, *488*, 275–290. [CrossRef]
26. Jackson, A.L.; Inger, R.; Parnell, A.C.; Bearhop, S. Comparing isotopic niche widths among and within communities: SIBER—Stable Isotope Bayesian Ellipses in R. *J. Anim. Ecol.* **2011**, *80*, 595–602. [CrossRef]
27. Altabet, M.A.; Pilskaln, C.; Thunell, R.; Pride, C.; Sigman, D.; Chavez, F.; Francois, R. The nitrogen isotope biogeochemistry of sinking particles from the margin of the Eastern North Pacific. *Deep. Sea Res. Part I Oceanogr. Res. Pap.* **1999**, *46*, 655–679. [CrossRef]
28. Graham, B.S.; Koch, P.L.; Newsome, S.D.; McMahon, K.W.; Aurioles, D. Using Isoscapes to Trace the Movements and Foraging Behavior of Top Predators in Oceanic Ecosystems. In *Isoscapes: Understanding Movement, Pattern, and Process on Earth through Isotope Mapping*; West, J., Bowen, G.J., Dawson, T.E., Tu, K.P., Eds.; Springer: Amsterdam, The Netherlands, 2009; pp. 299–318.
29. García-Pámanes, J.; Trasviña-Castro, A.; Lara-Lara, J.R.; Bazán-Guzmán, C. Variaciones estacionales del flujo vertical de materia orgánica particulada en la región central del Golfo de California. *Cienc. Mar.* **2011**, *37*, 33–49.
30. Guzman-del Proo, S.A.; Mille-Pagaza, S.R.; Campa-Guzmán, S.; Carrillo-Laguna, J.; Pereira Corona, A.; Belmar-Perez, J.; Parra-Alcocer, M.J.; Luque-Guerrero, A.C. La comunidad bentónica de los bancos de abulón (*Haliotis* spp. Mollusca: Gastropoda) en Bahía Tortugas, Baja California Sur, México. *An. Esc. Nac. Cienc. Biológicas* **1991**, *36*, 27–59.
31. Newsome, S.D.; Rio, C.M.; Bearhop, S.; Phillips, D.L. A niche for isotopic ecology. *Front. Ecol. Environ.* **2007**, *5*, 429–436. [CrossRef]
32. MacNeil, M.A.; Skomal, G.B.; Fisk, A.T. Stable isotopes from multiple tissues reveal diet switching in sharks. *Mar. Ecol. Prog. Ser.* **2005**, *302*, 199–206. [CrossRef]
33. Kim, S.L.; Casper, D.R.; Galván-Magaña, F.; Ochoa-Díaz, R.; HernándezAguilar, S.B.; Koch, P.L. Carbon and nitrogen discrimination factors for elasmobranch soft tissues based on a long-term controlled feeding study. *Environ. Biol. Fishes* **2012**, *95*, 37–52. [CrossRef]

diversity

MDPI

Article

Trophic Partitioning among Three Mesopredatory Shark Species Inhabiting the Northwestern Adriatic Sea

Licia Finotto [1], Daniela Berto [2], Federico Rampazzo [2], Saša Raicevich [2], Sara Bonanomi [3] and Carlotta Mazzoldi [1,4,*]

[1] Department of Biology, University of Padova, Via Ugo Bassi 58/B, 35121 Padova, Italy; licia.finotto@unipd.it
[2] Institute for Environmental Protection and Research (ISPRA), Loc. Brondolo, 30015 Venice, Italy; daniela.berto@isprambiente.it (D.B.); federico.rampazzo@isprambiente.it (F.R.); sasa.raicevich@isprambiente.it (S.R.)
[3] National Research Council (CNR), Institute of Marine Biological Resources and Biotechnologies (IRBIM), Largo Fiera della Pesca 1, 60125 Ancona, Italy; sara.bonanomi@cnr.it
[4] CoNISMa (Interuniversitary Consortium of Marine Sciences), Piazzale Flaminio 9, 00196 Roma, Italy
* Correspondence: carlotta.mazzoldi@unipd.it; Tel.: +049-8276194

Abstract: While the general diet of Mediterranean elasmobranchs has been widely studied, little is known about food partitioning and competition among sympatric species, despite these being important forces structuring marine communities. Using stomach content and stable isotope analyses, we investigated diet and trophic levels and evaluated the diet overlap and partitioning of *Scyliorhinus canicula*, *Mustelus mustelus*, and *M. punctulatus* in the northwestern Adriatic Sea. These shark species were confirmed as opportunistic mesopredators, but significant differences in their diets emerged. The two bentho-demersal *Mustelus* species had a larger trophic overlap with *S. canicula* than between each other. Given the pronounced morphological similarity of these two *Mustelus* species, this is likely a strategy to limit competition. The strictly benthic *S. canicula* showed a more varied diet compared to the other species. Stable isotope analysis highlighted that despite the smaller size and overlapping diets, *S. canicula* occupied a slightly higher trophic level. A better characterization of the trophic role of these species in the food web of the basin can be obtained from these data. At an ecosystem level, this information is essential to evaluate the possible consequences of the decline or recovery of the population of these exploited species.

Keywords: diet; competition; diet overlap; feeding habits; sympatric species; trophic level; stable isotope; *Scyliorhinus canicula*; *Mustelus mustelus*; *Mustelus punctulatus*

1. Introduction

Inter- and intra-specific competition represents an important force in structuring marine communities [1]. By reducing the pressure of competition, resource and food partitioning is the main process allowing for the coexistence of sympatric species or of different life stages of the same species [2,3]. Food partitioning is even more important when co-occurring predatory species have similar morphology (i.e., mouth shape, dentition, and body shape with the associated swimming abilities), habitat usage, and foraging habits (i.e., living and feeding either in the water column or close to the sea bottom). Indeed, similar characteristics allow predators to target and hunt the same prey species [2,4], increasing potential competition. This is particularly the case for congeneric species living in the same community given their low evolutionary divergence and similar specialization [5]. Partitioning can occur along different levels, such as time, space, life stages, and trophic niches [6–8]. A better characterization of trophic relationships, energy transfer, resource partitioning, and competition occurrence can improve our understanding of the structure, dynamics, and functioning of marine communities [6,9,10].

Elasmobranchs (sharks, skates, and rays) are key top and mesopredatory species whose predatory activity can be highly influential in marine ecosystems [11–13], regulating both fish and invertebrate populations [14]. Mesopredatory elasmobranchs are extremely important because they mediate the changes in community structure and functions caused by the loss of apex predators [15,16]; nevertheless, diet information is often lacking [17]. Despite many species, especially opportunistic feeders, showing a significant dietary geographical variation [18,19], data on diet and trophic relationships are often limited to only a few geographic areas and populations. Even fewer data exist on competition and resource partitioning between co-occurring elasmobranch species [20]. Improved knowledge of the trophic ecology of mesopredatory elasmobranchs, including diet composition, characterization of regional diets, niche breadth, and diet overlap between sympatric species, is necessary. Indeed, this information can provide a better understanding of elasmobranchs' functional role as top-down regulators in marine communities and of the processes of energy transfer in trophic webs and marine ecosystems. In this context, combining the two commonly used techniques of stomach content and stable isotope analysis is beneficial. This approach can overcome the drawbacks of each technique, i.e., a high number of samples required for stomach content analysis and the low specific and temporal resolution of stable isotope analysis, allowing for a better characterization of the diet and trophic role of the investigated species ([21–23], and references within).

The northern Adriatic Sea is one of the most productive [24,25] and fished sub-basins of the Mediterranean Sea [26,27]. Shallow waters and muddy–sandy bottoms characterize the Italian coasts, while deeper waters and rocky substrates typify the Slovenian and Croatian side [25]. This offers suitable habitats to a great variety of species, thus sustaining high biodiversity [26]. Several mesopredatory elasmobranchs co-occur in this region [26,28], many of which have been overfished in the past [26,29]. Among these, three of the most abundant shark species, *Mustelus mustelus*, *M. punctulatus*, and *Scyliorhinus canicula* [28], underwent a severe decline [26,29]. These species share similar benthopelagic habits and a diet composed of mainly crustaceans and, in smaller proportions, teleosts, mollusks, and polychaetes [30–32]. *S. canicula* is a small (up to 50.5 cm; [33]), benthic species that inhabits the continental shelf and uppermost slopes on rocky to sandy bottoms down to a depth of 400 m, where it rests in crevices and holes [30]; therefore, it is more abundant on the eastern side of the study area, which is characterized by rocky substrates [25]. This is an oviparous species that lays eggs throughout the year in the study area and reaches maturity at small sizes (around 40–41 cm; [33]). Due to its sedentary habits, *S. canicula* resides in the northwestern Adriatic Sea throughout the year [33], although some sexual segregation occurs in the study area, as also reported in other populations [33–35]. The two *Mustelus* species are large (up to 158 and 141 cm for *M. mustelus* and *M. punctulatus*, respectively; [36]) viviparous species that reach sexual maturity at large sizes (110–120 cm [36]) and share very similar morphology, life history traits [36,37], and habitats. These congeneric species have a demersal habit and spend most of the time swimming in midwater or, more commonly, near the bottom, down to a depth of 350 m [30]. Both species perform seasonal migrations in the study area, arriving when the water temperature starts to increase (April–May) and leaving when temperatures start to drop (November–December; [36]), moving to the southern part of the basin. The northwestern Adriatic Sea is used as a parturition and mating area, and, at different times, mature pregnant females, mature actively reproducing males, and juveniles and neonates of both sexes can be found [36].

The diet of these three species has been previously investigated in different areas [32,38–41], also including the northeastern Adriatic Sea [42–46]. Nevertheless, apart from a comparison between the two *Mustelus* species in the Strait of Sicily [31], no data exist on diet overlap and resource partitioning among these three species. Yet, their co-occurrence and similar morphology and habits suggest that a strong trophic competition might exist. Obtaining in-depth information on their trophic ecology is essential to understand their functional role in the ecosystem and the potential top-down consequences of the observed population decline [15,16]. In this context, this study aimed to (1) better characterize the diet and trophic strategy of these three species in the northwestern Adriatic Sea using stomach content and stable isotope analyses; (2) highlight intra-specific diet differences related to sex, size, or season; and (3) identify potential competition, resource partitioning, or diet overlap between the three species.

2. Materials and Methods

2.1. Stomach Content Analysis

Between 2012 and 2013, 480 specimens (Table 1) of *Scyliorhinus canicula* (N = 243), *Mustelus mustelus* (N = 114), and *M. punctulatus* (N = 123) were sampled from the landings of Chioggia's fishing fleet operating in the northwestern Adriatic Sea [26,47]. *S. canicula* was identified based on morphological features while, for *Mustelus* specimens, the species was genetically attributed following Marino et al. [37]. Sex was attributed according to the presence of claspers in males. Total length and body mass were recorded using a measuring tape (0.5 cm accuracy) and a scale (0.1 kg), respectively. Excised stomachs were preserved in 70% ethanol in seawater. After removing excess ethanol with tissue paper, the total mass of the stomach content was measured with a precision scale (0.01 g). Prey was identified using a stereomicroscope to the lowest possible taxonomic level according to the available identification keys [48–55], counted, and weighed (0.01 g). When only body parts were found, the smallest number of individuals from which the fragments could have originated was recorded. Unidentifiable material was not included in the analysis.

Table 1. Total length (cm; mean ± standard deviation) of *Scyliorhinus canicula*, *Mustelus mustelus*, and *M. punctulatus* used for stomach content (non-empty stomach) and stable isotope analyses. Data are presented and divided into sex, size, and season groups in *S. canicula* and for sex and size groups in *M. mustelus* and *M. punctulatus*. Numbers in parentheses represent the sample size of each group.

Stomach content analysis			***Scyliorhinus canicula***			
		Small		Big		Total
		Cold season	Hot season	Cold season	Warm season	
	Males	41.3 ± 0.4 (2)	35.3 ± 2.5 (3)	45.8 ± 2.7 (6)	45.5 ± 2.2 (31)	44.6 ± 3.5 (42)
	Females	40.3 ± 1.4 (11)	40.1 ± 1.9 (18)	44.6 ± 1.6 (70)	44.4 ± 1.8 (59)	43.7 ± 2.4 (158)
	Total	39.8 ± 2.2 (34)		44.8 ± 1.9 (166)		
			Mustelus mustelus			
		Small		Big		Total
	Males	62.0 ± 12.8 (21)		106.4 ± 10.0 (6)		71.9 ± 22.4 (27)
	Females	63.4 ± 13.3 (28)		118.7 ± 16.3 (47)		98.1 ± 30.9 (75)
	Total	62.8 ± 13.0 (49)		117.3 ± 16.2 (53)		
			Mustelus punctulatus			
		Small		Big		Total
	Males	39.0 ± 3.0 (49)		106.9 ± 11.0 (12)		52.4 ± 27.7 (61)
	Females	38.9 ± 5.0 (43)		108.9 ± 11.1 (9)		51.0 ± 27.5 (52)
	Total	39.0 ± 4.0 (92)		107.8 ± 10.8 (21)		
Stable isotope analysis			***Scyliorhinus canicula***			
		Small		Big		Total
	Males	/		45.2 ± 1.8 (12)		
	Females	/		45.3 ± 1.8 (12)		
	Total	/		45.2 ± 1.8 (24)		
			Mustelus mustelus			
		Small		Big		Total
	Males	73.0 ± 9.4 (4)		114.2 ± 14.4 (6)		97.7 ± 24.4 (10)
	Females	69.8 ± 11.5 (6)		125.5 (1)		77.8 ± 23.5 (7)
	Total	71.1 ± 10.3 (10)		115.8 ± 13.8 (7)		
			Mustelus punctulatus			
		Small		Big		Total
	Males	53.7 ± 6.4 (5)		101.0 ± 9.6 (9)		84.1 ± 25.0 (14)
	Females	50.5 (1)		105.1 ± 15.5 (9)		99.6 ± 22.6 (10)
	Total	53.2 ± 5.9 ()		103.0 ± 12.7 (18)		

To investigate whether a sufficient number of stomachs were analyzed, cumulative diversity curves were made using the lowest taxonomic level of the prey, separately for each species. Using EstimateS software (version 9.1) [56], the order in which the stomachs were analyzed was randomized 500 times, and the Shannon–Weaver index (H′, mean ± standard deviation; see below for its calculation), a proxy for diet diversity, was plotted against the total number of non-empty stomachs [57]. The presence of an asymptote in the curve indicates that enough stomachs were analyzed [58].

For each stomach, the percent fullness (%fullness), a proxy for feeding intensity [59], was calculated by dividing the total mass of the stomach content by the body mass of the specimen and multiplying by 100. Diet breadth was investigated with two diversity indices; the Shannon–Weaver index (H′; [60]) [61] was calculated as:

$$H' = \sum_{i=1}^{n} p_i \cdot \ln(p_i) \quad (1)$$

where p_i is the amount of prey category i (g) relative to the totality of the prey categories found in the stomach (n). The Pielou index is a proxy for diet evenness, highlighting the potential dominance of few prey categories in the diet (J′; [62]), and was computed as:

$$J' = H'/H_{max} \quad (2)$$

where H_{max} is the maximum value that H' can assume, equal to log(S), where S is the total number of prey categories found in the stomach.

For each prey category, the percent frequency of occurrence ($\%FO_i$), the prey-specific abundance ($\%PN_i$), and the prey-specific weight ($\%PW_i$) were calculated as:

$$\%FO_i = (n_i/N) \times 100 \quad (3)$$

where n_i is the number of stomachs in which prey category i was found, and N is the total number of non-empty stomachs.

$$\%PN_i = \left(\sum_{j=1}^{ni} N_{ij}\right)/n_i \quad (4)$$

where $\%N_{ij}$ is the numerical abundance of prey category I in stomach sample j, and n_i is the number of stomachs containing prey category i;

$$\%PW_i = \left(\sum_{j=1}^{ni} W_{ij}\right)/n_i \quad (5)$$

where $\%W_{ij}$ is the abundance by weight of prey category i in stomach sample j, and ni is the number of stomachs containing prey category i.

Using these indices, the prey-specific index of relative importance ($\%PSIRI_i$) was calculated as [63]:

$$\%PSIRI_i = 0.5 \times \%FO_i \times (\%PN_i + \%PW_i) \quad (6)$$

To investigate feeding strategies, $\%PN_i$ was plotted against $\%FO_i$ [64].

To check for the existence of trophic overlap among each predator pair combination, the simplified Morisita–Horn (M–H) index [65] was computed using $\%PW_i$:

$$M\text{–}H = \left(2\sum_{i=1}^{n} P_{ij}\, P_{ik}\right)/\left(\sum_{i=1}^{n} P_{ij}^{\,2} P_{ik}^{\,2}\right) \quad (7)$$

where n is the total number of prey categories, p_{ij} is the proportion of the prey category i consumed by predator j, and p_{ik} is the proportion of the prey category i consumed by predator k. According to the criteria proposed by Langton [66], an M–H value > 0.6 indicates a high dietary overlap, values ranging from 0.3 to 0.59 correspond to a medium overlap, and values < 0.29 indicate a low overlap.

2.2. Stable Isotope Analysis

From a representative subsample of specimens analyzed for the stomach content (Table 1), about 1 cm^3 of white muscle was excised from below the first dorsal fin. Additionally, in spring–summer of 2013, individuals of the dominant prey categories (teleosts, crustaceans, and mollusks) were collected from catches from otter-trawl vessels fishing in the northern Adriatic Sea (Table S4). At least three individuals for each prey category were processed to obtain muscle tissue (about 2 g). For the smallest prey, different samples were pooled and analyzed together.

Muscle samples were oven-dried at 60 °C for 48 h, then ground into a fine powder using a combusted mortar and pestle. Samples were not subjected to lipid extractions since they all had consistently low lipid content (carbon (C): nitrogen (N) < 4.0; [67]). The stable C and N ratios were measured using an Isotope Ratio Mass Spectrometer DeltaV Advantage (Thermo Fisher Scientific, Bremen, Germany) together with a CHN Analyzer Flash 2000 (Thermo Fisher Scientific, Bremen, Germany). The ratio of stable isotopes was expressed in delta (δ) notation:

$$\delta = [(\text{Rsample}/\text{Rstandard}) - 1)] \times 10^3 \tag{8}$$

where δ is the isotope ratio of the sample relative to the standards (international standard Vienna Pee Dee Belemnite (VPDB) for C and atmospheric nitrogen for N). Rsample is the fraction of heavy to light isotopes in the sample, while Rstandard is the fraction measured in the standard. The multiplication by 1000 is used to express the δ notation as units of parts per thousand (‰). An internal standard (mussel muscle) was analyzed throughout each run and was both accurate and precise (−20.5 ± 0.5‰ for δ^{13}C and 5.7 ± 0.2‰ for δ^{15}N). For both δ^{13}C and δ^{15}N, the analytical precision of measurements was 0.2%. Sucrose IAEA CH6 (International Atomic Energy Agency, Vienna, Austria), L-glutamic acid (RM 8574, National Institute of Standards and Technology, NIST, Gaithersburg, MA, USA), caffeine (National Institute of Standards and Technology, NIST, Gaithersburg, MA, USA), and urea (National Institute of Standards and Technology, NIST, Gaithersburg, MA, USA) were used as certified reference materials.

Using δ^{15}N, the trophic level of species was estimated according to Fortibuoni et al. [68] with an enrichment factor ΔN of 3.4‰. The δ^{15}N value of Adriatic zooplankton (6.6 ‰) used as a baseline for the primary consumer of trophic level one was assumed from Berto, D. (Institute for Environmental Protection and Research (ISPRA), Venice, Italy), unpublished data, 2023.

2.3. Data Analysis

To test differences in diet, three dichotomous factors were taken into consideration: season (warm or cold), sex, and size (small or big). We considered size rather than sexual maturity because, in elasmobranchs, size accounts for dietary ontogenetic shifts and predatory abilities [7,69]. The mean total length observed in the northern Adriatic Sea (males and females combined: 41.5, 92.9, and 76.1 cm for *S. canicula*, *M. mustelus*, and *M. punctulatus*, respectively; [33,36]) was used as a threshold to allocate individuals to the big- or the small-sized group. Species sampled during spring or summer (March–August) were allocated to the warm season group, while those sampled during autumn or winter (September–February) were allocated to the cold season group.

Before analysis, the normality and the homogeneity of variance of the datasets were checked with the Shapiro–Wilk test and Lavene's test, respectively. If, after transformation, data were not normally distributed, non-normal distributions or non-parametric tests were used. For a better description of the sample and for the interpretation of the results, separately for each species, chi-squared tests were used to investigate differences in the proportion of males and females between the two size groups and between the two seasons and to test differences in the proportion of big and small individuals between the two seasons. Potential differences in %fullness, H′, and J′ between sexes, size groups, and seasons were tested. A linear model was used, including sex, size, and their interaction as fixed factors. Due to the small sample size, a separate linear model was used to test the factor season. Before analysis, an arcsin of the square root transformation of %fullness data was used [70]; for clarity, results were reported as percentages.

A permutational multivariate analysis of variance (PERMANOVA), including sex, size, season, and their interaction, was used to highlight any intraspecific differences in diet [5,71–74]. The analysis was based on the Bray–Curtis similarity matrix obtained from the square root transformation of the prey biomass data. A non-metric multidimensional scaling (nMDS) was performed to graphically represent the dissimilarity between the different groups [5]. Prior to the analysis, prey categories with %FOi lower than 5% were aggregated in larger categories according to taxonomic and ecological criteria, with the exclusion of prey belonging to species of commercial values or showing high abundance in the diet of one of the species (Table S1). A PERMANOVA test including the factor species (three levels) was used to investigate any interspecific differences in diet composition. To identify differences between species pairs, pairwise PERMANOVA comparisons were performed. A similarity percentage (SIMPER) analysis based on square root transformed biomass data was used to identify the percent dissimilarities between the groups and the percent contribution of the different prey categories to the observed differences [5,71–74].

Given the lack of normality, potential differences in stable isotope values were investigated using non-parametric Kruskal–Wallis tests. Differences between size groups (big and small) within each species (except for *S. canicula*) and differences between the three species were tested separately.

Results were reported as mean ± standard deviation (S.D.). A significance level of $\alpha = 0.05$ was used for the tests. Data were analyzed using R statistical software (version 4.2.2), with lme4 [75], lsmeans [76], and emmeans [77] packages (R Core team 2021) and Statsoft (Ver. 5.0). Multivariate analyses were performed using PRIMER 6 and PERMANOVA+.

3. Results

Of the analyzed stomachs, 35 (14.4%), 12 (10.5%), and 10 (8.1%) contained only parasites (cestodes and nematodes) in *Scyliorhinus canicula*, *Mustelus mustelus*, and *M. punctulatus*, respectively. They were thus considered empty and not included in the analyses. A summary of the total length and of the number of males/females for each species, size group, and season is reported in Table 1 and Figure S1. In *M. mustelus*, the proportion of males and females in the big- and the small-sized groups significantly differed ($\chi^2 = 13.01$, $p < 0.001$), while it was similar in *S. canicula* ($\chi^2 = 0.98$, $p = 0.32$) and *M. punctulatus* ($\chi^2 = 0.10$, $p = 0.74$). In *S. canicula*, a significantly higher proportion of females was observed in the cold season ($\chi^2 = 13.95$, $p < 0.001$), while the proportion of small and big animals did not differ between seasons ($\chi^2 = 0.65$, $p = 0.42$). All the *M. punctulatus* specimens and all the big *M. mustelus* females were sampled only in the warm season; therefore, the factor season was excluded from the analyses for these two species.

The cumulative diversity curves reached an asymptote for the three species, indicating that the number of analyzed stomachs was high enough to describe their diet (Figure 1).

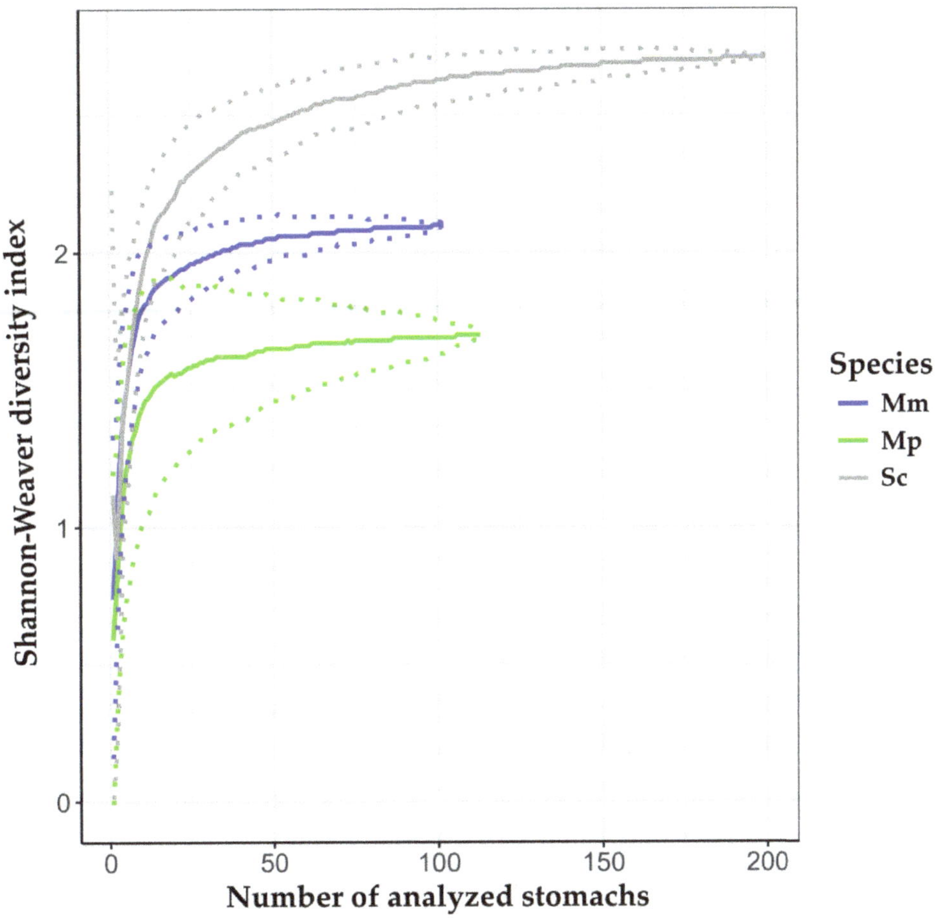

Figure 1. Shannon–Weaver diversity index (Mean, solid line ± standard deviation, dotted lines) for the cumulative diversity curve of *Scyliorhinus canicula* (SC, grey), *Mustelus mustelus* (Mm, blue), and *M. punctulatus* (Mp, green).

3.1. *Intraspecific Analysis*

3.1.1. Scyliorhinus canicula

The main taxonomic categories in the diets were crustaceans (%PSIRI = 59.6%) and teleost fishes (25.3%), followed by cephalopods (9.5%) and polychaetes (4.7%; Table S1). In particular, the crustaceans *Liocarcinus depurator*, unidentified Caridea, unidentified Portunidae, unidentified Brachiura, *Liocarcinus* sp., and *Rissoides desmaresti* were the most important prey categories (%PSIRI > 4%; Table S1). *S. canicula* had a generalist feeding strategy, as almost all prey categories were rare, being present in less than 25% of the analyzed stomachs, and all having a prey-specific abundance (%PN) lower than 50% (Figure 2a). Sex and size, but not their interaction, and season had a significant effect on %fullness; values were higher in females, in small animals, and in the warm season. Size, sex, and their interaction did not have a significant effect on the diversity indices, while H′ and J′ were both significantly higher in the warm season (Table 2). PERMANOVA highlighted significant trophic differences between sexes and seasons, but not between size groups (Table 3; Figure 3a,b). The average diet similarity within males and females was 15.3% and 18.5%, respectively, and the average dissimilarity between sexes was 86.2%. The average similarity within the warm and the cold season was 20.9% and 15.4%, respectively, while the average difference between the two seasons was 82.5%. Table 4 reports the main prey categories contributing to the observed differences.

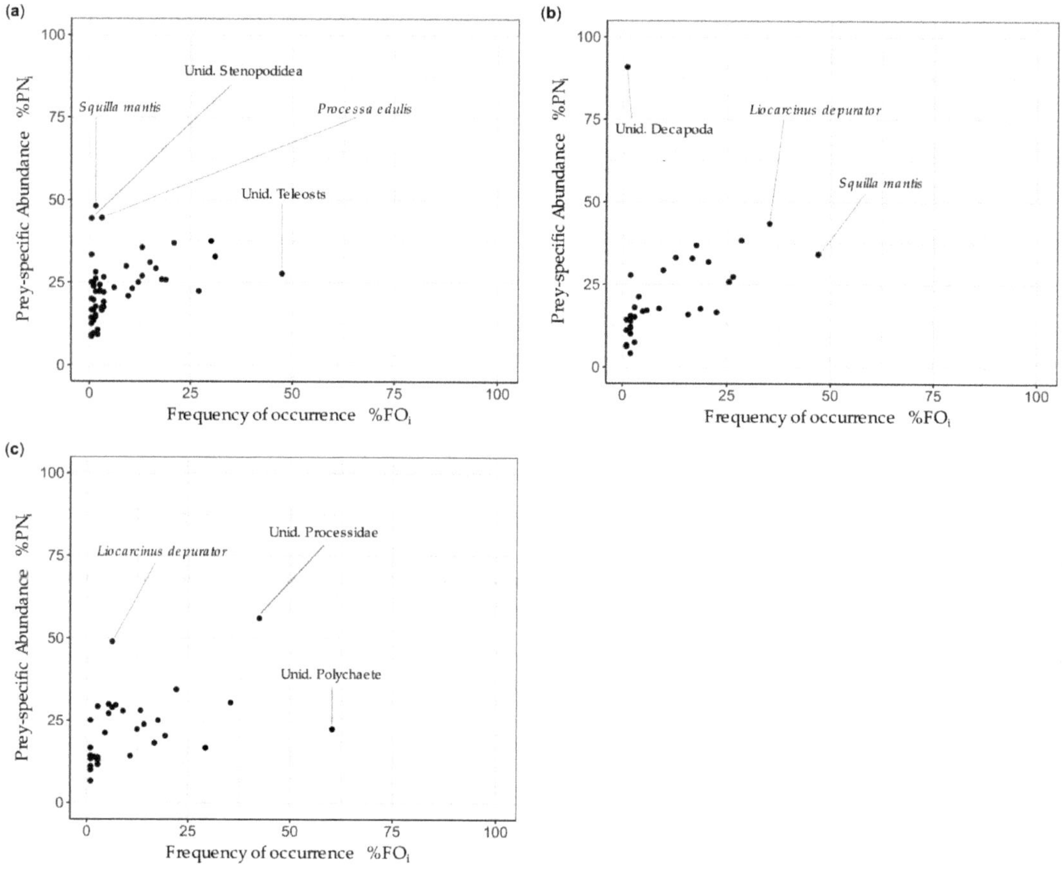

Figure 2. Feeding strategy of (**a**) *S. canicula*, (**b**) *M. mustelus*, and (**c**) *M. punctulatus* represented as the prey-specific abundance (%PN) of every prey category identified in the diet plotted against its percent frequency of occurrence (%FOi). Each black dot corresponds to a different prey category identified in the diet of the species.

Table 2. Results of the linear models testing the effect of size, sex, season, and the interaction between size and sex on fullness% and Shannon–Wiener and Pielou diversity indices for *S. canicula* (Sc), *M. mustelus* (Mm), and *M. punctulatus* (Mp). Statistically significant results are in bold. The mean (±standard deviation) of each group is reported.

	Fullness%			**Shannon–Wiener**			**Pielou**		
	Sc	**Mm**	**Mp**	**Sc**	**Mm**	**Mp**	**Sc**	**Mm**	**Mp**
Size	$F_{2,197} = 4.45$ **$p < 0.001$**	$F_{3,98} = 0.44$ $p = 0.66$	$F_{3,109} = 4.02$ **$p < 0.001$**	$F_{3,196} = 0.96$, $p = 0.34$	$F_{3,98} = 0.51$ $p = 0.61$	$F_{3,109} = 0.88$ $p = 0.38$	$F_{3,178} = 0.23$ $p = 0.82$	$F_{3,90} = 0.49$ $p = 0.62$	$F_{3,101} = 0.27$ $p = 0.79$
Big	1.90 ± 1.88%	1.29 ± 0.72%	0.68 ± 0.49%	0.70 ± 0.46	0.73 ± 0.41	0.85 ± 0.52	0.62 ± 0.26	0.64 ± 0.22	0.66 ± 0.24
Small	4.32 ± 3.47%	2.08 ± 1.41%	1.48 ± 0.92%	0.82 ± 0.42	0.70 ± 0.46	0.89 ± 0.47	0.63 ± 0.22	0.58 ± 0.26	0.75 ± 0.22

Table 2. *Cont.*

	Fullness%			Shannon–Wiener			Pielou		
	Sc	**Mm**	**Mp**	**Sc**	**Mm**	**Mp**	**Sc**	**Mm**	**Mp**
Sex	$F_{2,197} = 2.45$ $\mathbf{p = 0.02}$	$F_{3,98} = 0.05$ $p = 0.96$	$F_{3,109} = 1.64$ $p = 0.10$	$F_{3,196} = 0.62$, $p = 0.53$	$F_{3,98} = 0.71$ $p = 0.48$	$F_{3,109} = 0.91$ $p = 0.37$	$F_{3,178} = 1.28$ $p = 0.20$	$F_{3,90} = 0.33$ $p = 0.74$	$F_{3,101} = 0.07$ $p = 0.95$
Males	1.41 ± 1.47%	1.65 ± 1.16%	1.32 ± 0.88%	0.76 ± 0.45	0.70 ± 0.49	0.88 ± 0.46	0.72 ± 0.23	0.59 ± 0.28	0.74 ± 0.19
Females	2.55 ± 2.54%	1.68 ± 1.18%	1.35 ± 0.95%	0.71 ± 0.46	0.72 ± 0.41	0.89 ± 0.49	0.59 ± 0.26	0.61 ± 0.23	0.72 ± 0.26
Size × sex	$F_{3,196} = 1.28$ $p = 0.20$	$F_{3,98} = 0.16$ $p = 0.87$	$F_{3,109} = 1.81$ $p = 0.07$	$F_{3,196} = 0.64$ $p = 0.52$	$F_{3,98} = 0.82$ $p = 0.41$	$F_{3,109} = 1.03$ $p = 0.31$	$F_{3,178} = 0.05$ $p = 0.96$	$F_{3,90} = 0.41$ $p = 0.68$	$F_{3,101} = 0.01$ $p = 0.99$
Season	$F_{1,198} = 2.14$ $\mathbf{p = 0.03}$	/	/	$F_{1,198} = 2.15$ $\mathbf{p = 0.03}$	/	/	$F_{1,180} = 2.22$ $\mathbf{p = 0.06}$	/	/
Warm	2.33 ± 2.22%			0.84 ± 0.42			0.69 ± 0.21		
Cold	2.28 ± 2.62%			0.58 ± 0.47			0.53 ± 0.28		

Table 3. Results of the permutational multivariate analysis of variance (PERMANOVA) on the dietary composition by biomass of *S.canicula*, *M. mustelus*, and *M. punctulatus*. Bold values highlight statistical significance. df = degrees of freedom, SS = sum of squares, and MS = mean sum of squares.

	Source	df	SS	MS	Pseudo-F	P (perm)	Unique Perms
S. canicula	Size	1	3631.9	3631.9	1.008	0.420	999
	Sex	1	10131	10131	2.811	**0.003**	997
	Season	1	7303.9	7303.9	2.027	**0.034**	999
	Size × Sex	1	3867.8	3867.8	1.073	0.370	999
	Size × Season	1	4132.5	4132.5	1.147	0.319	999
	Sex × Season	1	3229.4	3229.4	0.896	0.571	999
	Size × Sex × Season	1	3468.8	3468.8	0.965	0.474	998
	Residual	192	6.9×10^5	3603.9			
	Total	199	7.4×10^5				
M. mustelus	**Size**	1	20353	20353	8.156	**0.001**	999
	Sex	1	1390.3	1390.3	0.557	0.737	999
	Size × Sex	1	6193.4	6193.4	2.482	**0.039**	999
	Residual	98	2.4×10^5	2495.5			
	Total	101	2.9×10^5				
M. punctulatus	**Size**	1	41462	41462	13.123	**0.001**	997
	Sex	1	3348.9	3348.9	1.060	0.391	999
	Size × Sex	1	4254.8	4254.8	1.347	0.204	996
	Residual	109	3.4×10^5	3159.5			
	Total	112	3.9×10^5				

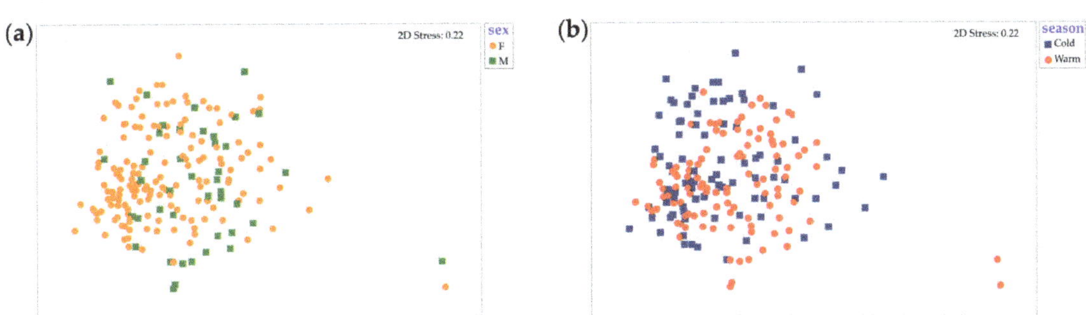

Figure 3. *Cont.*

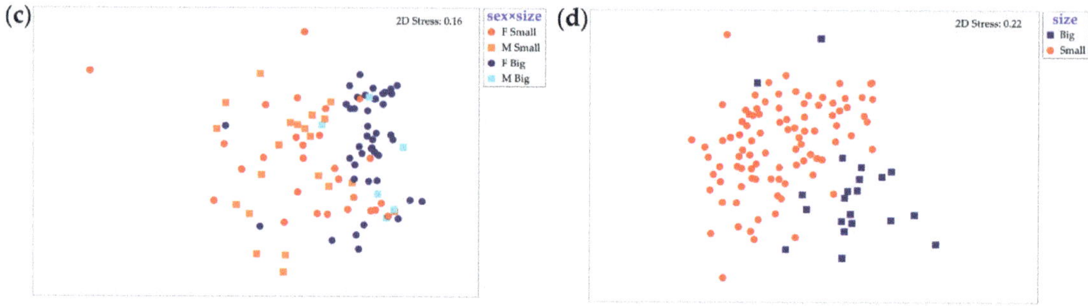

Figure 3. Non-metric multidimensional scaling (nMDS) ordination of dietary composition by biomass of (**a**) sexes (females, F: orange •; males, M: green ■) and (**b**) seasons in *S. canicula* (cold season, Cold: blue ■; warm season, Warm: red •), (**c**) the combination of the levels of sex and size factors in *M. mustelus* (small females, F Small: red •; small males, M Small: orange ■; big females, F Big: blue •; big males, M Big: light blue ■), and (**d**) sizes in *M. punctulatus* (small animals, Small: red •; big animals, Big: dark blue ■).

Table 4. Results of the similarity percentage (SIMPER) analysis reporting the average biomass and percentage contribution of the different prey categories to the difference observed between sex and season groups in the diet of *S. canicula*, and between size groups in the diet of *M. mustelus* and *M. punctulatus*. Only prey categories contributing at least 4% to the difference were included in the table.

Species	Factor	Prey Category	Average Biomass	Average Biomass	Contribution (%)
			Females	**Males**	
S. canicula	Sex	Portunidae	1.15	0.33	21.77
		Unid. Teleosts	0.44	0.24	9.24
		Other Caridea	0.14	0.33	7.45
		Unid. Crustaceans	0.19	0.22	6.62
		Sepiolidae	0.17	0.22	5.91
		Other Brachiura	0.18	0.19	5.82
		Rissoides desmaresti	0.11	0.14	4.63
		Cepola macrophthalma	0.15	0.12	4.36
		Alpheus glaber	0.08	0.17	4.19
		Flatfishes	0.29	0.06	4.10
			Cold season	**Warm season**	
	Season	Portunidae	0.95	0.99	24.02
		Unid. Teleosts	0.46	0.35	10.70
		Flatfishes	0.36	0.14	6.54
		Unid. Crustaceans	0.11	0.26	6.35
		Other Brachiura	0.16	0.20	5.46
		Sepiolidae	0.11	0.23	4.97
		Deltentosteus quadrimaculatus	0.22	0.11	4.62
		Other Caridea	0.14	0.22	4.61
		Cepola macrophthalma	0.09	0.19	4.31
		Unid. Polychaete	0.12	0.13	4.10
			Small animals	**Big animals**	
M. mustelus	Size	Portunidae	1.39	5.83	39.39
		Squilla mantis	1.40	2.59	21.43
		Other Brachiura	0.40	0.69	7.48
		Ethusa mascarone	0.85	0.00	6.68
		Unid. Crustaceans	0.39	0.25	4.24
		Other Stomatopoda	0.22	0.37	4.20

Table 4. *Cont.*

Species	Factor	Prey Category	Average Biomass	Average Biomass	Contribution (%)
M. punctulatus	Size	Portunidae	0.04	2.20	25.17
		Squilla mantis	0.02	0.85	9.13
		Unid. Teleosts	0.15	0.69	7.30
		Other Polychaete	0.40	0.49	6.41
		Pelagic fishes	0.13	0.48	5.45
		Other Brachiura	0.20	0.46	5.35
		Other	0.02	0.48	5.00
		Unid. Cephalopods	0.14	0.41	4.56
		Anomura	0.35	0.07	4.36
		Unid. Processidae	0.36	0.00	4.19

3.1.2. Mustelus mustelus

This diet was almost exclusively composed of crustaceans (%PSIRI = 95.3%). In particular, *Squilla mantis*, *L. depurator*, unidentified Portunidae, *Ethusa mascarone*, *Liocarcinus* sp., unidentified Brachiura, and *Carcinus aestuarii* were the most important prey categories (Table S1). Almost all prey categories were rare; however, *S. mantis* was present in more than 45% of the stomachs, and *L. depurator* and unidentified Decapoda had a prey-specific abundance higher than 40% (Figure 2b). This suggests that *M. mustelus* had mostly a generalist feeding strategy, apart from a weak specialization for the abovementioned prey categories. Size, sex, and their interaction did not have a significant effect on %fullness or the H′ or the J′ index (Table 2). PERMANOVA analysis highlighted the significant effects of the interaction between sex and size and of the factor size (Table 3 and Figure 3c); the pairwise test highlighted that small and big animals were significantly different from one another independently from the sex, but small females and big males did not differ (Table 5); however, only a few big males were sampled (N = 6). The average diet similarity within small and big animals was 25.0% and 42.3%, respectively, while the average dissimilarity between size groups was 77.6% (Table 4; see Table S2 in Supplementary Materials for the results of the interaction between sex and size).

Table 5. Results of the pairwise test investigating the effect of the interaction between the factors sex and size on the dietary composition by biomass of *M. mustelus*. Bold values highlight statistical significance. SF = small females, SM = small males, BF = big females, and BM = big males.

Comparison	*t*	P (perm)	Unique Perms
SF, SM	1.2579	0.140	998
SF, BF	3.3407	**0.001**	999
SF, BM	1.3271	0.120	998
SM, BF	3.4784	**0.001**	999
SM, BM	1.8145	**0.008**	995
BF, BM	1.2952	0.173	998

3.1.3. Mustelus punctulatus

This diet was dominated by crustaceans (66.2%), but polychaetes (14.1%), teleosts (9.0%), and cephalopods (8.7%) were also present. According to %PSIRI, unidentified Processidae, *Dardanus* sp., unidentified Caridea, and *Dardanus calidus* were the most important prey (Table S1). Almost all prey categories were rare, apart for unidentified polychaetes and unidentified Processidae, which were present in more than 60% and 40% of the stomachs, respectively, and *L. depurator* and unidentified Processidae, which had a prey-specific abundance approaching 50% and higher than 55%, respectively (Figure 2c). This suggests that *M. punctulatus* had a weak specialization for polychaetes, *L. depurator*, and Processidae, and otherwise, a mostly generalist feeding strategy. Size, but neither sex nor their interaction, had a significant effect on %fullness; values for small animals were significantly higher than those measured in big animals. Size, sex, and their interaction did not have a significant effect on H′ or J′ index (Table 2). PERMANOVA analysis highlighted a significant difference in the diet between size groups (Table 3, Figure 3d). The average similarity in the diet within small and big animals was 22.5% and 23.5%, respectively, while the average dissimilarities between size groups was 92.6% (Table 4).

3.2. *Interspecific Comparison*

The simplified Morisita–Horn index is equal to 0.71 between the pair *S. canicula* and *M. mustelus*, 0.54 between *S. canicula* and *M. punctulatus*, and 0.32 between *M. mustelus* and *M. punctulatus*. PERMANOVA analysis highlighted a significant difference in the diet between the three species, and the pairwise test showed that all species differed from each other (Figure 4; Table 6). The average diet similarity was 16.1% within *S. canicula*, 28.3% within *M. mustelus*, and 17.9% within *M. punctulatus*. The average dissimilarity was 87.4% between *S. canicula* and *M. mustelus*, 89.5% between *S. canicula* and *M. punctulatus*, and 91.1% between *M. mustelus* and *M. punctulatus* (Table 7).

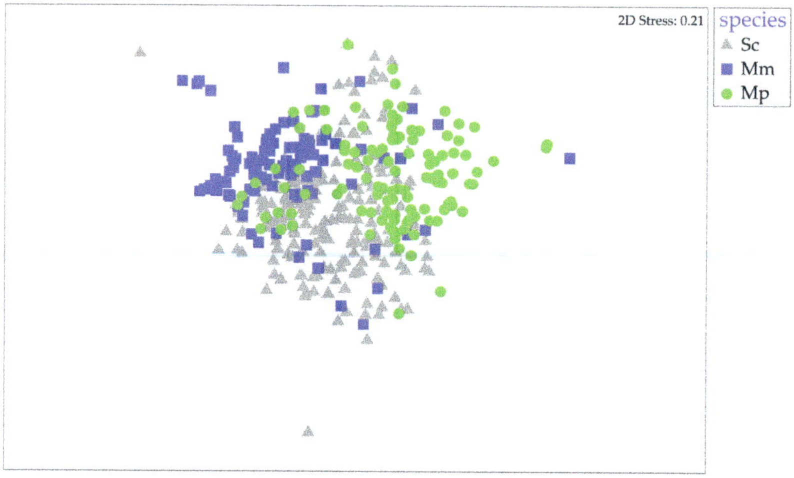

Figure 4. Non-metric multidimensional scaling (nMDS) ordination of dietary composition by biomass of *S. canicula* (Sc, grey ▲), *M. mustelus* (Mm, blue ■), and *M. punctulatus* (Mp, green •).

Table 6. Results of permutational multivariate analysis of variance (PERMANOVA) on the dietary composition by biomass of the three species and of the associated pairwise tests. Bold values highlight statistical significance. Sc = *S. canicula*, Mm = *M. mustelus*, and Mp = *M. punctulatus*. df = degrees of freedom, SS = sum of squares, and MS = mean sum of squares.

PERMANOVA						
Source	**df**	**SS**	**MS**	**Pseudo-F**	**P (perm)**	**Unique Perms**
Species	2	1.8×10^5	91733	26.52	**0.001**	998
Residual	412	1.4×10^5	3459.1			
Total	414	1.6×10^5				
	Pairwise comparison					
Pair-wise comparison	*t*	P (perm)	Unique perms			
Sc, Mm	5.211	**0.001**	999			
Sc, Mp	4.492	**0.001**	999			
Mm, Mp	6.002	**0.001**	999			

Table 7. Results of the similarity percentage (SIMPER) analysis reporting the average biomass and percentage contribution of the different prey categories to the difference observed in the diet between *S. canicula* and *M. mustelus*, *S. canicula* and *M. punctulatus*, and *M. mustelus* and *M. punctulatus*. Only prey categories contributing at least 4% to the difference were included in the table.

Prey Category	Average Biomass	Average Biomass	Contribution (%)
	S. canicula	***M. mustelus***	
Portunidae	0.97	3.70	31.01
Squilla mantis	0.01	2.02	18.80
Other Brachiura	0.18	0.55	6.75
Unid. Teleosts	0.40	0.27	5.24
Ethusa mascarone	0.00	0.41	5.17
Unid. Crustaceans	0.19	0.32	4.57
	S. canicula	***M. punctulatus***	
Portunidae	0.97	0.44	17.72
Unid. Teleosts	0.40	0.25	8.32
Other Polychaete	0.13	0.42	7.30
Unid. Crustaceans	0.19	0.25	6.60
Processidae	0.10	0.29	6.57
Anomura	0.01	0.30	6.00
Other Brachiura	0.18	0.25	5.75
Other Caridea	0.18	0.18	5.52
Sepiolidae	0.18	0.13	4.50
	M. mustelus	***M. punctulatus***	
Portunidae	3.70	0.44	30.86
Squilla mantis	2.02	0.17	18.62
Other Brachiura	0.55	0.25	6.77
Ethusa mascarone	0.41	0.11	5.52
Unid. Crustaceans	0.32	0.25	4.78
Other Polychaete	0.04	0.42	4.18

3.3. Stable Isotope

Variation in average $\delta^{15}N$ per species was restricted, ranging from 12.09‰ to 13.59‰; on the opposite, $\delta^{13}C$ values ranged from −13.13‰ to −17.60‰ (Figure 5, Table S3). $\delta^{15}N$ did not differ significantly among the three species, nor between big- and small-sized individuals for *M. mustelus* ($p > 0.05$), while for *M. punctulatus*, significant differences were observed ($H_{1,23} = 7.111$, $p = 0.008$). No significant differences were found between sexes within species (all $p > 0.05$). $\delta^{13}C$ showed high variation within species (Figure 5, Table S3), presenting significantly higher values (always above −16.00‰) in big individuals of *M. punctulatus* ($H_{1,21} = 6.788$, $p = 0.009$) and *M. mustelus* ($H_{1,15} = 9.070$, $p = 0.003$) as compared to the small ones. Pertaining to prey, a decreasing trophic position in the food web was observed from teleosts to crustaceans and mollusks (bivalves), with average $\delta^{15}N$ decreasing from 11.85‰ to 8.92‰ and 5.09‰, respectively, while $\delta^{13}C$ average values ranged from −17.00‰ to −19.20‰ (Figure 5, Table S4).

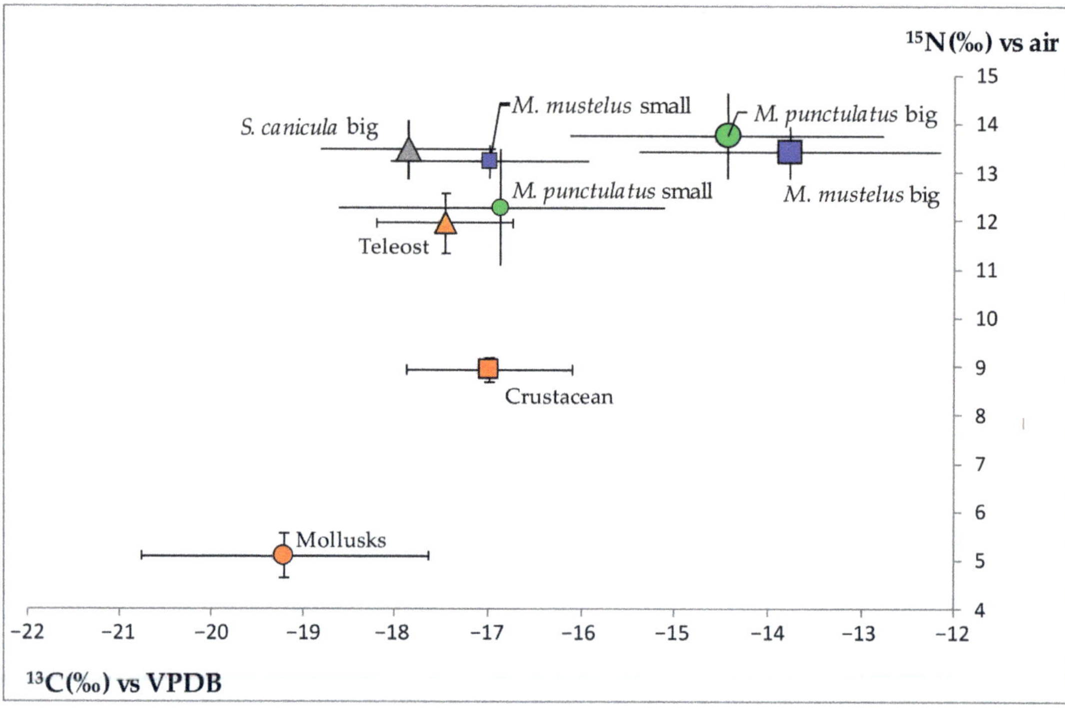

Figure 5. Bidimensional plot of carbon ($\delta^{13}C$) and nitrogen ($\delta^{15}N$) stable isotopes ratios (mean ± standard deviation) for the three shark species categorized for size (*S. canicula*: grey ▲; *M. mustelus*: blue ■; *M. punctulatus*: green •) and for their main prey categories (Teleost: orange ▲; Crustacean: orange ▲; Mollusks: orange ▲). Data are reported as the isotope ratio of the sample relative to the standards (international standard Vienna Pee Dee Belemnite (VPDB) for C and atmospheric nitrogen for N).

4. Discussion

To better characterize the processes structuring communities in the northwestern Adriatic Sea, we investigated the diet and trophic strategy of three of the most abundant shark species in the area, namely *Scyliorhinus canicula*, *Mustelus mustelus*, and *M. punctulatus*. For the first time for the area, we also assessed their trophic positions via stable isotope analysis and investigated their potential diet overlaps and resource partitioning.

This study confirms the important role of the three studied species in the northern Adriatic Sea as mesopredators [31,78] feeding on crustaceans, teleosts, mollusks, and polychaetes. Moreover, the three species were confirmed as generalist predators [79], although the two *Mustelus* species showed some weak specializations for some prey categories. Despite the general common preponderance of crustaceans in the diet of the three species, some dissimilarities emerged from the comparisons with dietary studies conducted in the Atlantic Ocean [80–87], other areas of the Mediterranean Sea [31,32,40,41,88,89], and even the northeastern Adriatic Sea [42–46,90]. This pronounced geographic variability likely results from the opportunistic foraging strategy of these species, whose diet reflects prey availability in different areas [18,19,32,40,87]. For the northern Adriatic Sea, differences in habitats between the two sides of the basin may account for the reported diet variability. Indeed, the importance of polychaetes in the diet that we observed in the northwestern Adriatic Sea for *S. canicula* and, especially, *M. punctulatus* can be attributed to the preponderance of muddy habitats in comparison with the rocky substrates of the eastern side [25].

Feeding intensity (i.e., stomach fullness; [59]) was higher in small *S. canicula* and *M. punctulatus* [45], and in female *S. canicula*. It is likely that both groups require higher amounts of energetic resources: young, small individuals to sustain faster growth rates [91,92], and females to sustain reproduction [32,34]. Stomach fullness was also higher in the warm season in *S. canicula*, potentially to sustain the faster metabolism associated with higher temperatures [93,94]. *S. canicula* diet was more diverse and homogeneous in the warm season, likely because of seasonal differences in prey availability [41]. As also reported in other areas [32], the diet of *S. canicula* differed between sexes. This species shows sexual segregation [34,35], with males and females occupying different habitats and, therefore, feeding on different prey. Alternatively, the sexual dimorphism of teeth and mouth morphology can explain the sexual dietary difference. Males have longer, sharper teeth that are more efficient at capturing soft-bodied prey [95], and indeed, a greater importance of Sepiolidae in the diet of male *S. canicula* is observed. Contrarily to previous studies [32,34,46], no ontogenetic shifts in diet were highlighted in *S. canicula*, possibly because of the narrow size range and low number of small animals sampled. Broadening the investigated size ranges could allow for highlighting patterns related to ontogenetic shifts in diet. On the other hand, ontogenetic diet shifts were confirmed in the two *Mustelus* species [31,41,44,88]. Smaller crustaceans (*Ethusa mascarone* and Anomura, mostly *Dardanus* sp. and Processidae) were more or exclusively present in the diet of small animals, while larger (Portunidae, *S. mantis*, and Brachiura) and faster prey (teleosts and cephalopods) were more abundant in the diet of big animals. In aquatic environments, predation is limited by the mouth gap [96], and as an individual grows, it is able to prey upon larger animals [41] thanks to a greater crushing capacity [87] and stronger bite and suction force [97]. Moreover, as animals grow, their swimming abilities also improve, allowing them to hunt faster animals [98]. In the northwestern Adriatic Sea, big *Mustelus* co-occur with smaller ones [26,36,37]; therefore, ontogenetic diet shifts are essential in reducing competition, allowing co-occurrence, and increasing survival and fitness [2,3,7].

Although the investigated species are generalist predators, their diets significantly differed. The important crustacean composition of the *Mustelus* species' diet [31,38,42,44] conforms to their dentition, presenting molariform teeth with weak cusps fused together at the base, creating a strong plate that can efficiently crush crustacean shells [30]. On the other hand, *S. canicula* has sharp teeth with two to four lateral cusps, also suitable for holding soft-bodied prey such as teleosts and mollusks [95]. Moreover, *S. canicula* is a small, benthic, sedentary species that lives mainly on rocky bottoms [30], while the two *Mustelus* species are larger [36], benthopelagic, active species. These marked ecological and morphological dissimilarities may explain the dietary differences observed between *S. canicula* and the two *Mustelus* species. On the other hand, *M. mustelus* and *M. punctulatus* co-occur in the same macro-geographic areas, exploit similar habitats, and have overlapping bathymetric distribution [26,99] and very similar morphology [37]; nonetheless, these two species differ in their diet. *M. mustelus* attains larger sizes [36] and can hunt larger and/or faster prey [96,98,100], as suggested by the higher importance of large crustaceans (Portunidae, *S. mantis*, and Brachiura) in its diet. Nevertheless, the diets of these two species also differed when larger *M. mustelus* were excluded from the analysis (Figure S2 and Tables S5 and S6), and therefore, size difference cannot be the only explanation. Despite similar morphology, the mouth and dermal denticles' shape vary between the two species [37], and preliminary observations suggest that body morphology also differs, including the shape and dimension of the first dorsal, pectoral, and caudal fins [101,102]. Fins are fundamental in determining swimming type, maneuverability, and performance [103–105]. Similarly, specific dermal denticles' morphologies are associated with the enhancement of swimming performance through drag reduction [106,107]. The difference existing in dermal denticles and fin morphology, along with those in mouth shape [108], possibly accounting for different swimming performances and predatory and feeding abilities, could be responsible for the observed diet dissimilarities. This would allow the two species to target different prey and finely differentiate the occupied habitat within the same broad geographic area and benthopelagic zone [103–105], as observed in other sympatric *Mustelus* species [87]. *M. mustelus*, consuming almost exclusively benthic crustaceans, seems to feed in close proximity to the seabed. On the other hand, *M. punctulatus* seems to also exploit the water column immediately above the sea bottom, as its diet is mostly composed of benthic prey but also benthopelagic teleost and mollusks.

The observed differentiated diets fit with those predicted by ecological theories for sympatric species; indeed, some degree of differentiation in the diets of species living in the same geographic area and habitat is functional to ensure resource partitioning and reduce the intensity of competition [2,3,7,109,110]. This mechanism could be the basis of the lower similarities between the two congeneric species respective to those with *S. canicula*. Contrarily, *S. canicula* has a broader diet, foraging opportunistically on all the variety of prey present in the rocky-bottom habitat it occupies [111], converging in the use of food resources with both *Mustelus* species without entering in strong competition with them [2,3,7,109]. Indeed, the diet similarity within *S. canicula* samples is the lowest observed among the three species, with the highest total number of prey taxa identified (Table S1). Both *M. mustelus* and *M. punctulatus* showed some, albeit weak, specializations in prey categories, while *S. canicula* showed none. All these observations seem to support the broad generalist foraging of *S. canicula*.

The results obtained from stomach content analysis are confirmed by isotope analysis, even considering the relatively limited number of data, especially for prey. The $\delta^{15}N$ values of the three species did not reveal significant differences in the trophic position [21–23]. In addition, no change for this parameter was observed when size and sex were considered, except for *M. punctulactus*, and only for size. While this result shows that the three species occupy the same trophic level, $\delta^{13}C$ highlights that this condition is mirrored by differences in terms of carbon sources in their diet [21–23]. In fact, for this parameter, marked and significant differences were observed between small and big individuals of the *Mustelus* species, a result that is consistent with the ontogenetic shift observed in these species through stomach content analyses. The direction of such change indicates a shift toward less negative values of $\delta^{13}C$ of large-sized specimens. Similar results were reported by Espinoza et al. [112] for other elasmobranch species, in which size was a relevant driver for the changes in $\delta^{13}C$ associated with differences in diet composition.

In terms of absolute values, the $\delta^{15}N$ observed in the muscle of the three shark species is compatible with the prevalence of crustaceans in their diet and a possible increase of about 3.0–3.4‰ with an increase of one trophic level [113]. However, the relevant prey categories of *S. canicula* and *M. punctulatus* also included other taxa, as shown by the stomach data. In this context, the still relatively low proportion of prey presenting either a higher trophic level (such as teleosts) or lower trophic level species (like bivalves) as compared to intermediate levels (crustaceans) could have prevented the emergence of clear differences in the trophic position of the three species. Such a pattern could have also been influenced by the fact that, as reported by Fortibuoni et al. [68], the $\delta^{15}N$ of elasmobranchs may present lower values as compared to the actual trophic position of the species due to high levels of urea retained by sharks for osmoregulatory purposes. In terms of $\delta^{13}C$, *S. canicula* and small *Mustelus* specimens showed values similar to those observed in their main prey, while larger *M. mustelus* and *M. punctulatus* presented higher $\delta^{13}C$ values. The higher $\delta^{13}C$ estimates for the big-sized individuals reflect an increase of the lighter ^{12}C isotope with respect to the heavier ^{13}C, likely favored by dietary adjustments determined by several factors, such as habitat changes, movement patterns, and predatory capacity. In this regard, it is likely that large-sized *Mustelus* specimens present a habitat use where prey have a higher incidence of carbon source related to marine productivity processes as compared to the coastal, inshore ones, which are characterized by lower (i.e., more negative) $\delta^{13}C$ values [21,22]. Therefore, large-sized specimens have a prevalence of food intake from sources of carbon from the open sea, possibly reflecting an ontogenetic shift in habitat.

Other factors that were not investigated could influence the diet, isotope values, and trophic relationships observed in this study. Isotopic values integrate diet over long periods [21–23]; because *Mustelus* species undertake seasonal migrations [36], their isotopic values could be influenced by the prey consumed during the period spent outside the study area. On the other hand, as stomach content analysis reflects more closely the prey consumed in a short period of time [21–23], the results obtained from this technique reliably represent the diet of the species in the study area. However, the results of the stomach content analysis accord with those of the stable isotope analysis, suggesting that these species also feed on prey with similar trophic levels outside the study area. Nevertheless, broadening the spatial scale of the assessment could further allow for distinguishing patterns related to changes in habitat use. The different bentho-demersal communities present in the habitats occupied by *S. canicula* and the *Mustelus* species could also have influenced the observed trophic partitioning [18,19,32,40,87]. However, refined information on small-scale distributions of prey species is not yet available. Finally, other mesopredatory bentho-demersal elasmobranchs could compete with the investigated species and influence the trophic relationships in the area. Indeed, several shark (*S. stellaris* and *Squalus acanthias*), skate (*Raja clavata* and *R. asterias*), and ray species (*Pteroplatytrygon violacea, Dasyatis pastinaca, Myliobatis aquila,* and *Aetomylaeus bovinus*) are abundant in the northwestern Adriatic Sea [26,28], and some of them, at least in other areas, show similar diets to those of the species investigated in this study [74,114–117]. Improving the knowledge of the diet of these species in the area will better characterize the trophic and community dynamics of the basin.

5. Conclusions

Our study confirmed the important mesopredatory role of the three investigated shark species, simultaneously highlighting some unexpected resource partitioning and, at the same time, trophic similarities between them. Given that most of the elasmobranch species living in the northern Adriatic Sea are exposed to high fishing pressures [26] and have undergone severe declines [26,29], further research is needed. Ultimately, these data should be used in ecosystem models aimed at investigating potential top-down consequences of the decline of these species [26].

Supplementary Materials: The following supporting information can be downloaded at: https://www.mdpi.com/article/10.3390/d15121163/s1, Figure S1: Length frequency distribution of the samples for each species; Figure S2: Non-metric multidimensional scaling ordination of dietary composition by biomass of small sized *M. mustelus* and *M. punctulatus*; Table S1: List of all prey categories from the stomachs of the three elasmobranch species; Table S2: Results of the SIMPER analysis in *M. mustelus*; Table S3: Carbon and nitrogen stable isotopes ratios; Table S4: Carbon and nitrogen stable isotopes ratios and estimated trophic value of the main prey items; Table S5: Results of permutational multivariate analysis of variance on the dietary composition by biomass of small sized *M. mustelus* and *M. punctulatus*. Table S6. Results of the SIMPER analysis in small sized *M. mustelus* and *M. punctulatus*.

Author Contributions: Conceptualization: L.F., C.M. and D.B.; Methodology: L.F., D.B. and F.R.; Data collection: L.F., D.B. and F.R.; Data analysis: L.F., D.B., F.R. and S.R.; Investigation: L.F., D.B. and F.R.; Writing—original draft preparation: L.F.; Writing—review and editing: D.B., F.R., S.B., S.R. and C.M.; Supervision: C.M.; Funding acquisition: C.M. and D.B. All authors have read and agreed to the published version of the manuscript.

Funding: This research was supported by funding from the Ministero dell'Istruzione, dell'Università e della Ricerca (MIUR) to CM.

Institutional Review Board Statement: Ethical review and approval were not required for this study since all the specimens were obtained from the catches of a commercial fishery and were not sacrificed to achieve the purposes of this study.

Data Availability Statement: Publicly available datasets were analyzed in this study. This data can be found in the figshare repository at: https://doi.org/10.6084/m9.figshare.24299407.

Acknowledgments: We wish to thank all the fishers who provided samples, as well as F. Badalamenti for advice in the sample collection.

Conflicts of Interest: The authors declare no conflict of interest.

References

1. Krebs, C.J. *Ecology*; Benjamin Cummings: San Francisco, CA, USA, 2001.
2. Langeland, A.; L'abée-lund, J.H.; Jonsson, B.; Jonsson, N. Resource partitioning and niche shift in Arctic charr *Savelinus alpinus* and brown trout *Salmo trutta*. *J. Anim. Ecol.* **1991**, *60*, 895–912. [CrossRef]
3. Ross, S.T. Resource partitioning in fish assemblages: A review of field studies. *Copeia* **1986**, *1986*, 352–388. [CrossRef]
4. Sibbing, F.A.; Nagelkerke, L.A.J. Resource partitioning by Lake Tana barbs predicted from fish morphometrics and prey characteristics. *Rev. Fish Biol. Fish.* **2001**, *10*, 393–437. [CrossRef]
5. Fanelli, E.; Badalamenti, F.; D'Anna, G.; Pipitone, C.; Riginella, E.; Azzurro, E. Food partitioning and diet temporal variation in two coexisting sparids, *Pagellus erythrinus* and *Pagellus acarne*. *J. Fish Biol.* **2011**, *78*, 869–900. [CrossRef] [PubMed]
6. Bornatowski, H.; Navia, A.F.; Braga, R.R.; Abilhoa, V.; Corrêa, M.F.M. Ecological importance of sharks and rays in a structural foodweb analysis in southern Brazil. *ICES J. Mar. Sci.* **2014**, *71*, 1586–1592. [CrossRef]
7. Papastamatiou, Y.P.; Wetherbee, B.M.; Lowe, C.J.; Crow, G.L. Distribution and diet of four species of carcharhinid shark in the Hawaiian Islands: Evidence for resource partitioning and competitive exclusion. *Mar. Ecol. Prog. Ser.* **2006**, *320*, 239–251. [CrossRef]
8. White, W.T.; Platell, M.E.; Potter, I.C. Comparisons between the diets of four abundant species of elasmobranchs in a subtropical embayment: Implications for resource partitioning. *Mar. Biol.* **2004**, *144*, 439–448. [CrossRef]
9. Bizzarro, J.J.; Carlisle, A.B.; Smith, W.D.; Cortés, E. Diet composition and trophic ecology of Northeast Pacific Ocean sharks. *Adv. Mar. Biol.* **2017**, *77*, 111–148.
10. Navia, A.F.; Cortés, E.; Mejía-Falla, P. Topological analysis of the ecological importance of elasmobranch fishes: A food web study on the Gulf of Tortugas, Colombia. *Ecol. Model.* **2010**, *221*, 2918–2926. [CrossRef]
11. Cortés, E. Standardized diet compositions and trophic levels of sharks. *J. Mar. Sci.* **1999**, *56*, 707–717. [CrossRef]
12. Heithaus, M.R. Predator–prey interactions. In *Biology of Sharks and Their Relatives*; Carrier, C., Musick, J.A., Heithaus, M.R., Eds.; CRC Press: Boca Raton, FL, USA, 2004.
13. Lucas, Z.; Stobo, W.T. Shark inflicted mortality on a population of harbour seals (*Phoca vitulina*) at Sable Island, Nova Scotia. *J. Zool.* **2000**, *252*, 405–414. [CrossRef]
14. Ellis, J.R.; Pawson, M.G.; Shackley, S.E. The comparative feeding ecology of six species of shark and four species of ray (Elasmobranchii) in the North-East Atlantic. *J. Mar. Biol. Assoc. UK* **1996**, *76*, 89–106. [CrossRef]
15. Ferretti, F.; Worm, B.; Britten, G.L.; Heithaus, M.R.; Lotze, H.K. Patterns and ecosystem consequences of shark declines in the ocean. *Ecol. Lett.* **2010**, *13*, 1055–1071. [CrossRef] [PubMed]
16. Ritchie, E.G.; Johnson, C.N. Predator interactions, mesopredator release and biodiversity conservation. *Ecol. Lett.* **2009**, *12*, 982–998. [CrossRef]
17. Jacobsen, I.P.; Bennett, M.B. A comparative analysis of feeding and trophic level ecology in stingrays (Rajiformes: Myliobatoidei) and electric rays (Rajiformes: Torpedinoidei). *PLoS ONE* **2013**, *8*, e71348. [CrossRef]
18. Ebert, D.A.; Cowley, P.D. Diet, feeding behaviour and habitat utilisation of the blue stingray *Dasyatis chrysonata* (Smith, 1828) in south African waters. *Mar. Freshw. Res.* **2003**, *54*, 957–965. [CrossRef]
19. Munroe, S.E.M.; Heupel, M.R.; Aaron, T.; Fisk, A.; Simpfendorfer, C.A. Geographic and temporal variation in the trophic ecology of a small bodied shark: Evidence of resilience to environmental change. *Can. J. Fish. Aquat. Sci.* **2015**, *72*, 343–351. [CrossRef]
20. Valls, M.; Quetglas, A.; Ordines, F.; Moranta, J. Feeding ecology of demersal elasmobranchs from the shelf and slope off the Balearic Sea (western Mediterranean). *Sci. Mar.* **2011**, *75*, 633–639. [CrossRef]
21. Chipps, S.R.; Garvey, J.E. Assessment of food habits and feeding patterns. In *Analysis and Interpretation of Freshwater Fisheries Data*; Guy, C.S., Brown, M.L., Eds.; American Fisheries Society: Bethesda, MD, USA, 2007.
22. Shiffman, D.S.; Gallagher, A.J.; Boyle, M.D.; Hammerschlag-Peyer, C.M.; Hammerschlag, N. Stable isotope analysis as a tool for elasmobranch conservation research: A primer for non-specialists. *Mar. Freshw. Res.* **2012**, *63*, 635–643. [CrossRef]
23. Barría, C.; Coll, M.; Navarro, J. Unravelling the ecological role and trophic relationships of uncommon and threatened elasmobranchs in the western Mediterranean Sea. *Mar. Ecol. Prog. Ser.* **2015**, *539*, 225–240. [CrossRef]
24. Fonda Umani, S. Pelagic production and biomass in the Adriatic Sea. *Sci. Mar.* **1996**, *60*, 6–77.
25. Giani, M.; Djakovac, T.; Degobbis, D.; Cozzi, S.; Solidoro, C.; Fonda Umani, S. Recent changes in the marine ecosystems of the northern Adriatic Sea. *Estuar. Coast. Shelf Sci.* **2012**, *115*, 1–13. [CrossRef]
26. Barausse, A.; Correale, V.; Curkovic, A.; Finotto, L.; Riginella, E.; Visentin, E.; Mazzoldi, C. The role of fisheries and the environment in driving the decline of elasmobranchs in the Northern Adriatic Sea. *ICES J. Mar. Sci.* **2014**, *71*, 1593–1603. [CrossRef]
27. Fortibuoni, T.; Giovanardi, O.; Pranovi, F.; Raicevich, S.; Solidoro, C.; Libralato, S. Analysis of long-term changes in a Mediterranean marine ecosystem based on fishery landings. *Front. Mar. Sci.* **2017**, *4*, 33. [CrossRef]
28. Relini, G.; Mannini, A.; De Ranieri, S.; Bitetto, I.; Follesa, M.C.; Gancitano, V.; Manfredi, C.; Casciaro, L.; Sion, L. Chondrichthyes caught during the MEDITS surveys in Italian waters. *Biol. Mar. Mediterr.* **2010**, *17*, 186–204.
29. Ferretti, F.; Osio, G.C.; Jenkins, C.J.; Rosenberg, A.A.; Lotze, H.K. Long-term change in a meso-predator community in response to prolonged and heterogeneous human impact. *Sci. Rep. Cetacean Res.* **2013**, *3*, 1057. [CrossRef] [PubMed]
30. Compagno, L.J.V. *Sharks of the World: An Annotated and Illustrated Catalogue of Shark Species Known to Date, Part 2, Carcharhiniformes*; FAO: Rome, Italy, 1984.

31. Di Lorenzo, M.; Vizzini, S.; Signa, G.; Andolina, C.; Palo, G.B.; Gristina, M.; Mazzoldi, C.; Colloca, F. Ontogenetic trophic segregation between two threatened smoothhound sharks in the Central Mediterranean Sea. *Sci. Rep.* **2020**, *10*, 11011. [CrossRef]
32. Kousteni, V.; Karachle, P.K.; Megalofonou, P. Diet of the small-spotted catshark *Scyliorhinus canicula* in the Aegean Sea (eastern Mediterranean). *Mar. Biol. Res.* **2017**, *13*, 161–173. [CrossRef]
33. Finotto, L.; Gristina, M.; Garofalo, G.; Riginella, E.; Mazzoldi, C. Contrasting life history and reproductive traits in two populations of *Scyliorhinus canicula*. *Mar. Biol.* **2015**, *162*, 1175–1186. [CrossRef]
34. Rodríguez-Cabello, C.; Sánchez, F.; Olaso, I. Distribution patterns and sexual segregations of *Scyliorhinus canicula* (L.) in the Cantabrian Sea. *J. Fish Biol.* **2007**, *70*, 1568–1586. [CrossRef]
35. Sims, D.W. Differences in habitat selection and reproductive strategies of male and female sharks. In *Sexual Segregation in Vertebrates: Ecology of the Two Sexes*; Ruckstuhl, K.E., Neuhaus, P., Eds.; Cambridge University Press: Cambridge, UK, 2005; p. 488.
36. Riginella, E.; Correale, V.; Marino, I.A.M.; Rasotto, M.B.; Vrbatovic, A.; Zane, L.; Mazzoldi, C. Contrasting life-history traits of two sympatric smooth-hound species: Implication for vulnerability. *J. Fish Biol.* **2020**, *96*, 853–857. [CrossRef]
37. Marino, I.A.M.; Finotto, L.; Colloca, F.; Di Lorenzo, M.; Gristina, M.; Farrell, E.D.; Zane, L.; Mazzoldi, C. Resolving the ambiguities in the identification of two smooth-hound sharks (*Mustelus mustelus* and *Mustelus punctulatus*) using genetics and morphology. *Mar. Biodivers.* **2018**, *48*, 1551–1562. [CrossRef]
38. Filiz, H. Diet composition of smooth-hound, *Mustelus mustelus* (Linnaeus, 1758), in Aegean Sea, Turkey. *Belg. J. Zool.* **2009**, *139*, 81–84.
39. Gravino, F.; Dimech, M.; Schembri, P.J. Feeding Habits of the Small-Spotted Catshark Scyliorhinus Canicula (L. 1758) in the Central Mediterranean. Paper presented at the Rapport du Congrès de la Commission Internationale pour l'Exploration Scientifique de la Mer Méditerranée. *CIESM Meet. Rapp. Comm. Int. Mer Médit.* **2010**, *39*, 1.
40. Saïdi, B.; Bradaï, M.N.; Bouaïn, A. Reproductive biology and diet of *Mustelus punctulatus* (Risso, 1826) (Chondrichthyes: Triakidae) from the Gulf of Gabès, central Mediterranean Sea. *Sci. Mar.* **2009**, *73*, 249–258. [CrossRef]
41. Saïdi, B.; Enajjar, S.; Bradaï, M.N.; Bouaïn, A. Diet composition of smooth-hound shark, *Mustelus mustelus* (Linnaeus, 1758), in the Gulf of Gabès, southern Tunisia. *J. Appl. Ichthyol.* **2009**, *25*, 113–118. [CrossRef]
42. Gračan, R.; Mladineo, I.; Lazar, B. Insight into the diet composition and gastrointestinal parasite community of the common smooth-hound, *Mustelus mustelus* (Carcharhiniformes: Triakidae), in the northern Adriatic Sea. *Nat. Croat.* **2014**, *23*, 35–44.
43. Jardas, I.; Šantić, M.; Nerlović, N.; Pallaoro, A. Diet composition of blackspotted smooth-hound, *Mustelus punctulatus* (Risso, 1826), in the eastern Adriatic Sea. *J. Appl. Ichthyol.* **2007**, *23*, 279–281. [CrossRef]
44. Jardas, I.; Šantić, M.; Nerlović, N.; Pallaoro, A. Diet of the smooth-hound, *Mustelus mustelus* (Chondrichthyes: Triakidae), in the eastern Adriatic Sea. *Cybium Int. J. Ichthyol.* **2007**, *31*, 459–464.
45. Lipej, L.; Mavrič, B.; Rešek, S.; Chérif, M.; Capapé, C. Food and feeding habits of the blackspotted smooth-hound, *Mustelus punctulatus* (Elasmobranchii: Carcharhiniformes: Triakidae), from the northern Adriatic. *Acta Ichthyol. Piscat.* **2011**, *41*, 171–177. [CrossRef]
46. Šantić, M.; Rađa, B.; Pallaoro, A. Feeding habits of small-spotted catshark (*Scyliorhinus canicula* Linnaeus, 1758) from the eastern central Adriatic Sea. *Mar. Biol. Res.* **2012**, *8*, 1003–1011. [CrossRef]
47. Barausse, A.; Michieli, A.; Riginella, E.; Palmeri, L.; Mazzoldi, C. Long-term changes in community composition and lifehistory traits in a highly exploited basin (northern Adriatic Sea): The role of environment and anthropogenic pressures. *J. Fish Biol.* **2011**, *79*, 1453–1486. [CrossRef]
48. AFORO. Available online: http://aforo.cmima.csic.es/index.jsp (accessed on 14 July 2023).
49. Falciai, L.; Minervini, R. *Guida dei Crostacei Decapodi d'Europa*; Franco Muzzio Editore: Padova, Italy, 1992; p. 282.
50. Fischer, W.; Bauchot, M.L.; Schneider, M. *Fiches FAO D'identification des Espèces Pour les Besoins de la Pêche. Méditerranée et Mer Noire. Zone de Pêche 37 (Vol. Volume I. Végétaux et Invertébrés)*; FAO: Rome, Italy, 1987.
51. Louisy, P. *Guida all'identificazione dei Pesci Marini d'Europa e del Mediterraneo. (Ed. italiana a cura di Trainito E.)*; Il Castello: Perugia, Italy, 2006.
52. Riedl, R. *Fauna e Flora del Mediterraneo*; Franco Muzzio Editore: Padova, Italy, 1991.
53. Šoljan, T. *I pesci dell'Adriatico*; Arnoldo Mondadori Editore: Verona, Italy, 1975.
54. Tortonese, E.; Lanza, B. *Piccola Fauna Italiana: Pesci, Anfibi e Rettili*; Martello Editore: Milano, Italy, 1968.
55. Whitehead, P.J.P.; Bauchot, M.L.; Hureau, J.C.; Nielsen, J.; Tortonese, E. *Fishes of the North-Eastern Atlantic and Mediterranean (Vol. 1, 2, 3)*; UNESCO: Paris, France, 1986.
56. Colwell, R.K.; EstimateS: Statistical Estimation of Species Richness and Shared Species from Samples. Version 9. Available online: http://purl.oclc.org/estimates (accessed on 14 July 2023).
57. Bizzarro, J.J.; Robison, H.J.; Rinewalt, C.S.; Ebert, D.A. Comparative feeding ecology of four sympatric skate species off central California, USA. *Environ. Biol. Fishes* **2007**, *80*, 197–220. [CrossRef]
58. Ferry, L.A.; Cailliet, G.M. Sample Size and Data Analysis: Are We Characterizing and Comparing Diet Properly? In Proceedings of the Feeding Ecology and Nutrition in Fish, Symposium Proceedings, San Francisco, CA, USA, 14–18 July 1996.
59. Hyslop, E.J. Stomach contents analysis—A review of methods and their application. *J. Fish Biol.* **1980**, *17*, 411–429. [CrossRef]
60. Shannon, C.E.; Weaver, W. *The Mathematical Theory of Communication*; Urbans, University of Illionis Press: Champaign, IL, USA, 1949.
61. Fanelli, E.; Badalamenti, F.; D'Anna, G.; Pipitone, C.; Romano, C. Trophodynamic effects of trawling on the feeding ecology of pandora, *Pagellus erythrinus*, off the northern Sicily coast (Mediterranean Sea). *Mar. Freshw. Res.* **2010**, *61*, 408–417. [CrossRef]

62. Pielou, E.C. The measurement of diversity in different types of biological collections. *J. Theor. Biol.* **1966**, *13*, 131–144. [CrossRef]
63. Brown, S.C.; Bizzarro, J.J.; Cailliet, G.M.; Ebert, D.A. Breaking with tradition: Redefining measures for diet description with a case study of the Aleutian skate *Bathyraja aleutica* (Gilbert 1896). *Environ. Biol. Fishes* **2012**, *95*, 3–20. [CrossRef]
64. Amundsen, P.A.; Gabler, H.M.; Staldvik, F.J. A new approach to graphical analysis of feeding strategy from stomach contents data—Modification of the Costello (1990) method. *J. Fish Biol.* **1996**, *48*, 607–614. [CrossRef]
65. Krebs, C.J. *Ecological Methodology*; Harper & Row Publishers: New York, NY, USA, 1989.
66. Langton, R.S. Diet overlap between the Atlantic cod, *Gadus morhua*, silver hake, *Merluccius bilinearis* and fifteen other northwest Atlantic finfish. *Fish. Bull.-Natl. Ocean. Atmos. Adm.* **1982**, *80*, 745–759.
67. Ehrich, D.; Tarroux, A.; Stien, J.; Lecomte, N.; Killengreen, S.; Berteaux, D.; Yoccoz, N.G. Stable isotope analysis: Modelling lipid normalization for muscle and eggs from arctic mammals and birds. *Methods Ecol. Evol.* **2011**, *2*, 66–76. [CrossRef]
68. Fortibuoni, T.; Noventa, S.; Rampazzo, F.; Gion, C.; Formalewicz, M.; Berto, D.; Raicevich, S. Evidences of butyltins biomagnification along the Northern Adriatic food web (Mediterranean Sea) elucidated by stable isotopes ratios. *Environ. Sci. Technol.* **2013**, *47*, 3370–3377. [CrossRef] [PubMed]
69. Simpfendorfer, C.A.; Goodreid, A.B.; McAuley, R.B. Size, sex and geographic variation in the diet of the tiger shark, *Galeocerdo cuvier*, from Western Australian waters. *Environ. Biol. Fishes* **2001**, *61*, 37–46. [CrossRef]
70. Gotelli, N.J.; Ellison, A.M. *A Primer of Ecological Statistics*, 2nd ed.; Sinauer Associates, Inc.: Sunderland, MA, USA, 2004.
71. Yemisken, E.; Forero, M.G.; Megalofonou, P.; Eryilmaz, L.; Navarro, J. Feeding habits of three Batoids in the Levantine Sea (north-eastern Mediterranean Sea) based on stomach content and isotopic data. *J. Mar. Biol. Assoc. UK* **2017**, *98*, 89–96. [CrossRef]
72. Pagliarini, C.D.; da Silva Riberiro, C.; Spada, L.; Delariva, R.L.; Araújo Chagas, J.M.; dos Anjos, L.A.; Ramos, I.P. Trophic ecology and metabolism of two species of nonnative freshwater stingrays (Chondrichthyes: Potamotrygonidae). *Hydrobiologia* **2020**, *847*, 2895–2908. [CrossRef]
73. Gonzalez-Pestana, A.; Silva-Garay, L.; Quiñones, J.; Mayaute, L.; Manrique, M.; Segura-Cobeña, E.; Espinoza, P.; Moscoso, V.; Vélez-Zuazo, X.; Alfaro-Shigueto, J.; et al. Geographic and ontogenetic variation in the diet of two commonly exploited batoids (Chilean eagle ray and Pacific guitarfish) off Peru: Evidence of trophic plasticity. *Environ. Biol. Fishes* **2021**, *104*, 1525–1540. [CrossRef]
74. Fanelli, E.; Da Ros, Z.; Martino, I.; Azzurro, E.; Bargione, G.; Donato, F.; Lucchetti, A. Crowding in the middle of marine food webs: A focus on *Raja asterias* and other mediterranean batoids. *Mar. Environ. Res.* **2022**, *183*, 105830. [CrossRef] [PubMed]
75. Bates, D.; Maechler, M.; Bolker, B.; Walker, S. Fitting linear mixed-effects models using lme4. *J. Stat. Softw.* **2015**, *67*, 1–48. [CrossRef]
76. Lenth, R.V. Least-Squares Means: The R Package lsmeans. *J. Stat. Softw.* **2016**, *69*, 1–33. [CrossRef]
77. Lenth, R.V. Emmeans: Estimated Marginal Means, aka Least-Squares Means. Available online: https://CRAN.R-project.org/package=emmeans (accessed on 14 July 2023).
78. Wieczorek, A.M.; Power, A.M.; Browne, P.; Graham, C.T. Stable-isotope analysis reveals the importance of soft-bodied prey in the diet of lesser spotted dogfish *Scyliorhinus canicula*. *J. Fish Biol.* **2017**, *93*, 685–693. [CrossRef]
79. Olaso, I.; Sánchez, F.; Rodríguez-Cabello, C.; Velasco, F. The feeding behaviour of some demersal fish species in response to artificial discarding. *Sci. Mar.* **2002**, *66*, 301–311. [CrossRef]
80. Eales, N.B. The food of the dogfish, *Scyliorhinus caniculus* L. *J. Mar. Biol. Assoc. UK* **1949**, *28*, 791–793. [CrossRef]
81. Henderson, A.C.; Dunne, J. Food of the lesser-spotted dogfish *Scyliorhinus canicula* (L.) in Galway Bay. *Ir. Nat. J.* **1999**, *26*, 191–194.
82. Martinho, F.; Sá, C.; Falcão, J.; Cabral, H.N.; Pardal, M.A. Comparative feeding ecology of two elasmobranch species, *Squalus blainville* and *Scyliorhinus canicula*, off the coast of Portugal. *Fish. Bull.* **2012**, *110*, 71–84.
83. Olaso, I.; Velasco, F.; Pérez, N. Importance of discarded blue whiting (*Micromesistius poutassou*) in the diet of lesser spotted dogfish (*Scyliorhinus canicula*) in the Cantabrian Sea. *ICES J. Mar. Sci.* **1998**, *55*, 331–341. [CrossRef]
84. Olaso, I.; Velasco, F.; Sánchez, F.; Serrano, A.; Rodríguez-Cabello, C.; Cendrero, D. Trophic relations of lesser-spotted catshark (*Scyliorhinus canicula*) and blackmouth catshark (*Galeus melastomus*) in the Cantabrian Sea. *J. Northwest Atl. Fish. Sci.* **2005**, *36*, 481–494. [CrossRef]
85. Saldanha, L.; Almeida, A.J.; Andrade, F.; Guerreiro, J. Observations on the diet of some slope dwelling fishes of southern Portugal. *Int. Rev. Der Gesamten Hydrobiol. Hydrogr.* **1995**, *80*, 217–234. [CrossRef]
86. Serrano, A.; Velasco, F.; Olaso, I.; Sánchez, F. Macrobenthic crustaceans in the diet of demersal fish in the Bay of Biscay in relation to abundance in the environment. *Sarsia* **2003**, *88*, 36–48. [CrossRef]
87. Smale, M.J.; Compagno, L.J.V. Life history and diet of two southern African smoothhound sharks, *Mustelus mustelus* (Linnaeus, 1758) and *Mustelus palumbes* Smith, 1957 (Pisces: Triakidae). *S. Afr. J. Mar. Sci.* **1997**, *18*, 229–248. [CrossRef]
88. Kara, A.F.; Al Hajaji, M.; Ghmati, H.; Shakman, E.A. Food and feeding habits of Mustelus mustelus (Linnaeus, 1758) (Chondrichthyes: Triakidae) along the Western coast of Libya. *Ann. Ser. Hist. Nat.* **2019**, *29*, 197–204. [CrossRef]
89. Karachle, P.K.; Stergiou, K.I. Food and Feeding Habits of Nine Elasmobranch Species in the Aegean Sea. *Rapp. Comm. Int. Mer. Médit.* **2010**, *39*, 553.
90. Jardas, I. Supplement to the knowledge of ecology of some Adriatic cartilaginous fishes (Chondrichthyes) with special reference to nutrition. *Act. Adriat.* **1972**, *14*, 1–60.

91. Boscolo Palo, G.; Di Lorenzo, M.; Gancitano, S.; Ragonese, S.; Mazzoldi, C.; Colloca, F. Sharks do not always grow slowly: Tagging data reveal a different pattern of growth, longevity and maturity for threatened smooth-hounds in the central Mediterranean Sea. *J. Mar. Sci. Eng.* **2022**, *10*, 1647. [CrossRef]
92. Ivory, P.; Jeal, F.; Nolan, C.P. Age determination, growth and reproduction in the lesser-spotted dogfish, *Scyliorhinus canicula* (L.). *J. Northwest Atl. Fish. Sci.* **2004**, *35*, 89–106. [CrossRef]
93. Chabot, D.; Steffensen, J.F.; Farrell, A.P. The determination of standard metabolic rate in fishes. *J. Fish Biol.* **2016**, *88*, 81–121. [CrossRef]
94. Whitney, N.M.; Lear, K.O.; Gaskins, L.C.; Gleiss, A.C. The effects of temperature and swimming speed on the metabolic rate of the nurse shark (*Ginglymostoma cirratum*, Bonaterre). *J. Exp. Mar. Biol. Ecol.* **2016**, *477*, 40–46. [CrossRef]
95. Ellis, J.R.; Shackley, S.E. Ontogenic changes and sexual dimorphism in the head, mouth and teeth of the lesser spotted dogfish. *J. Fish Biol.* **1995**, *47*, 155–164. [CrossRef]
96. Wilga, C.D.; Motta, P.J.; Sanford, C.P. Evolution and ecology of feeding in elasmobranchs. *Int. Comp. Biol.* **2007**, *47*, 55–69. [CrossRef]
97. Kolmann, M.A.; Huber, D.R.; Motta, P.J.; Grubbs, R.D. Feeding biomechanics of the cownose ray, *Rhinoptera bonasus*, over ontogeny. *J. Anat.* **2015**, *227*, 341–351. [CrossRef]
98. Kemper, J.M.; Bizzarro, J.J.; Ebert, D.A. Dietary variability in two common Alaskan skates (*Bathyraja interrupta* and *Raja rhina*). *Mar. Biol.* **2017**, *164*, 52. [CrossRef]
99. Colloca, F.; Enea, M.; Ragonese, S.; Di Lorenzo, M. A century of fishery data documenting the collapse of smooth-hounds (*Mustelus* spp.) in the Mediterranean Sea. *Aquat. Conserv. Mar. Freshw.* **2017**, *27*, 1145–1155. [CrossRef]
100. Herrel, A.; Gibb, A.C. Ontogeny of performance in vertebrates. *Physiol. Biochem. Zool.* **2006**, *79*, 1–6. [CrossRef]
101. Bressan, G. Studio Morfologico Della Pinna Caudale di Trentatré Specie di Squali (Elasmobranchii). Bachelor's Thesis, University of Padova, Padova, Italy, 2019.
102. Scarpa, A. Differenze Morfologiche fra i Palombi Mustelus Mustelus (Linnaeus, 1758) e Mustelus Punctulatus Risso, 1827 (Elasmobranchii, Triakidae). Bachelor's Thesis, University of Padova, Padova, Italy, 2017.
103. Deynat, P. *Les Requins. Identification des Nageoires. Guide Pratique*; Editions Quae: Versailles, France, 2010.
104. Irschick, D.J.; Fu, A.; Lauder, G.V.; Wilga, C.D.; Kuo, C.Y.; Hammerschlag, N. A comparative morphological analysis of body and fin shape for eight shark species. *Biol. J. Linn. Soc.* **2017**, *122*, 589–604. [CrossRef]
105. Lipej, L.; De Maddalena, A.; Soldo, A. *Sharks of the Adriatic Sea*; Knjiznica Annales Majora: Koper, Slovenia, 2004.
106. Lauder, G.V.; Wainwright, D.K.; Domel, A.G.; Weaver, J.C.; Wen, L.; Bertoldi, K. Structure, biomimetics, and fluid dynamics of fish skin surfaces. *Physic. Rev. Fluids* **2016**, *1*, 060502. [CrossRef]
107. Raschi, W.; Tabit, C. Functional aspects of placoid scales: A review and update. *Aust. J. Mar. Freshw. Res.* **1992**, *43*, 123–147. [CrossRef]
108. Rodríguez-Mendoza, R.; Muñoz, M.; Saborido-Rey, F. Ontogenetic allometry of the bluemouth, *Helicolenus dactylopterus dactylopterus* (Teleostei: Scorpaenidae), in the Northeast Atlantic and Mediterranean based on geometric morphometrics. *Hydrobiologia* **2011**, *670*, 5–22. [CrossRef]
109. Wetherbee, B.M.; Cortés, E.; Bizzarro, J.J. Food consumption and feeding habits. In *Biology of Sharks and their Relatives*; Carrier, J.C., Musick, J.A., Heithaus, M.R., Eds.; CRC Press: Boca Raton, FL, USA, 2012; pp. 239–263.
110. Baje, L.; Chin, A.; White, W.T.; Simpfendorfer, C.A. Dietary overlap of carcharhinid sharks in the Gulf of Papua. *Mar. Freshw. Res.* **2022**, *73*, 605–614. [CrossRef]
111. Ross, S.T. Patterns of resource partitioning in searobins (Pisces: Triglidae). *Copeia* **1977**, *1977*, 561–571. [CrossRef]
112. Espinoza, M.; Munroe, S.E.M.; Clarke, T.M.; Fisk, A.T.; Wehrtmann, I.S. Feeding ecology of common demersal elasmobranch species in the Pacific coast of Costa Rica inferred from stable isotope and stomach content analyses. *J. Exp. Mar. Biol. Ecol.* **2015**, *470*, 12–25. [CrossRef]
113. Post, D.M. Using Stable Isotopes to Estimate Trophic Position: Models, Methods, and Assumptions. *Ecology* **2002**, *83*, 703–771. [CrossRef]
114. Demirhan, S.A.; Seyhan, K.; Başusta, N. Dietary overlap in spiny dogfish (*Squalus acanthias*) and thornback ray (*Raja clavata*) in the southeastern Black Sea. *Ekoloji* **2007**, *16*, 1–8.
115. Šantić, M.; Rada, B.; Pallaoro, A. Diet and feeding strategy of thornback ray *Raja clavata*. *J. Fish Biol.* **2012**, *81*, 1070–1084. [CrossRef] [PubMed]
116. Lipej, L.; Mavrič, B.; Paliska, D.; Capapé, C. Feeding habits of the pelagic stingray *Pteroplatytrygon violacea* (Chondrichthyes: Dasyatidae) in the Adriatic Sea. *J. Mar. Biol. Assoc. UK* **2013**, *93*, 285–290. [CrossRef]
117. Kadri, H.; Marouani, S.; Bradai, M.N.; Bouaïn, A. Diet and feeding strategy of thornback ray, *Raja clavata* (Chondrichthyes; Rajidae) from the Gulf of Gabes (Tunisia–Central Mediterranean Sea). *J. Mar. Biol. Assoc. UK* **2014**, *94*, 1509–1516. [CrossRef]

diversity

MDPI

Review

A Review of the Paleobiology of Some Neogene Sharks and the Fossil Records of Extant Shark Species

Olaf Höltke *, Erin E. Maxwell and Michael W. Rasser

Staatliches Museum für Naturkunde Stuttgart, Rosenstein 1, 70191 Stuttgart, Germany; erin.maxwell@smns-bw.de (E.E.M.); michael.rasser@smns-bw.de (M.W.R.)
* Correspondence: olaf.hoeltke@smns-bw.de

Abstract: In recent years, new findings and new methods (stable isotopes of oxygen, zinc, and nitrogen; 2D and 3D modeling; and geometric morphometric analyses of the teeth) have enhanced our knowledge of the Neogene shark fauna and its paleobiology. Several papers deal with the large *Otodus* (*Megaselachus*) species, including the construction of a 3D model, as well as insights into its lifestyle and diet. In addition, the skeletal remains of *Carcharias gustrowensis*, *Carcharodon hastalis*, and *Keasius parvus* and a natural tooth set of *Carcharodon hubbelli* have been described in the last 13 years, and the dentition of the Neogene species *Carcharoides catticus*, *Megachasma applegatei*, and *Parotodus benedenii* has been reconstructed. Stable isotope analyses of the teeth from the Neogene species of *Araloselachus*, *Carcharias*, *Carcharodon*, *Galeocerdo*, *Hemipristris*, and *Mitsukurina* have given insights into the trophic positions of these genera during the Neogene, and shark teeth preserved near the skeletal remains of prey animals (mammals) and shark bite traces on these remains provide direct evidence of trophic interactions. The tooth shape, fossil locality, and paleoenvironment have been used to better understand the taxa *Carcharhinus dicelmai*, *Megalolamna paradoxodon*, *Pachyscyllium dachiardii*, and *P. distans*. Among extant species, *Galeorhinus galeus* can be traced back to the Eocene. *Alopias superciliosus*, *Rhincodon typus*, and possibly *A. vulpinus* can be traced back to the Oligocene. Species present by the Miocene include *Alopias vulpinus*, *Carcharhinus amblyrhynchoides*, *C. amblyrhynchos*, *C. albimarginatus*, *C. amboinensis*, *C. brachyurus*, *C. brevipinna*, *C. falciformis*, *C. glaucus*, *C. leucas*, *C. limbatus*, *C. longimanus*, *C. macloti*, *C. obscurus*, *C. perezi*, *C. sealei*, *Centrophorus granulosus*, *Cetorhinus maximus*, *Dalatias licha*, *Deania calcea*, *Galeocerdo cuvier*, *Glyphis glyphis*, *Heptranchias perlo*, *Isurus paucus*, *Lamna nasus*, *Negaprion brevirostris*, *Odontaspis ferox*, *Pseudocarcharias kamoharai*, *Sphyrna media*, *S. mokarran*, and possibly *Carcharodon carcharias*. First appearing in the Pliocene are *Scymnodon ringens*, *Somniosus rostratus*, and *Zameus squamulosus*. For some extant species (*Carcharias taurus*, *Hexanchus griseus*, *Isurus oxyrinchus*, *Notorynchus cepedianus*, and *Sphyrna zygaena*), it is not clear whether the assigned Neogene teeth represent the same species. The application of new methods to more fossil shark taxa, a detailed search for shark fossils, and better knowledge of the dentition of extant species (especially those with minute-sized teeth) will further enhance our knowledge of the evolution and paleobiology of sharks.

Keywords: Selachii; Miocene; Pliocene; paleobiology; ecology; Recent; *megalodon*

1. Introduction

The earliest record of elasmobranch fishes is probably from isolated scales potentially referable to the chondrichthyans, which date back to the Late Ordovician Epoch, about 455 million years ago [1]. Apart from a different tooth shape, Paleozoic sharks had a different anatomy from "modern" sharks (Neoselachii), which are known from the beginning of the Mesozoic era. Four key differences separating neoselacians from Paleozoic sharks were mentioned [2,3]: The jaws of neoselachians open wider than in earlier forms because of the greater mobility in the jaw joint and a highly kinetic palatoquadrate and hyomandibular. The notochord is enclosed in and constricted by calcified cartilaginous vertebrae, whereas primitive chondrichthyans had a simple notochordal sheath. The limb girdles in neoselachians are strengthened by fusion or firm connection on the midline, which allows for more powerful muscle activity. The basal elements (the radials) in the paired fins are reduced, and most of the fin is supported by flexible collagenous rods called ceratotrichia or actinotrichia. However, the general phylogeny, synapomorphies, evolution, and origin of elasmobranchs (sharks and rays) and holocephalans are still under discussion [4–7]. The rise and diversification of Neoselachii began in the Early Triassic Epoch, and, by the Neogene period, the shark fauna was similar to the Recent fauna. However, despite general similarities, the timing of the appearance of extant morphospecies, the extinction of some Paleogene–Neogene species, and the potential trophic changes resulting from these origin and extinction dynamics can provide insights into the structure and occupancy of higher trophic levels in Recent oceans.

The cartilaginous skeleton of sharks is normally not preserved in the fossil record, making the teeth the most abundant records of fossil sharks. Sharks replace their teeth continuously throughout their lifetime, and this high production of potential bioclasts makes fossil shark teeth the main vertebrate fossils in marine deposits of the Paleogene and Neogene periods. Therefore, the designation of species is mostly based on a few isolated teeth. In some cases, calcified vertebral centra can be found, as well as dermal denticles, fin spines, and gill rakers. The skeleton, or parts of it, were only fossilized under specific environmental conditions (e.g., fast sedimentation and exclusion of oxygen). Accordingly, such findings are very rare [8–11].

The "classical" method to infer the shark ecology from teeth is to look to extant relatives as analogues, as well as the shape of the teeth. The teeth were divided the different tooth shapes into eight adaptive dental types [3]. In addition to the tooth size and shape, the embedding sediment also gives an indication of the habitat preferences of Neogene sharks. In the last 20 to 30 years, new findings, as well as new methods, have made it possible to obtain more detailed information on the paleoecology of Neogene sharks. Recently, It was quantified the classical method by applying 2D geometric morphometrics to statistically discriminate the diet based on tooth shape, and it was also determined that variations in tooth morphology could be partitioned into seven key variables with which ecological roles in fossil sharks could be accurately assessed [12,13]. Paleobiology is probably best documented for the most famous fossil shark, *Otodus* (*Megaselachus*) *megalodon*, simply because there have been so many recent papers with this species as the main subject. The aim of this paper is to provide a detailed overview of those Neogene shark species for which the most data are available, excluding taxa described from only one or a few teeth. We then summarize what is known of the paleobiology of these Neogene shark species, as well as examining the first appearance of Recent species in the Neogene period (or sometimes earlier). The classification is based on Cappetta [3]. Genera and species are arranged in alphabetical order within higher taxonomic groupings. Lastly, we provide an outlook on possible future developments concerning research on fossil sharks. This work presents the current state of the art concerning the paleobiology of Neogene sharks, as well as the fossil records of extant species.

2. Methods Used to Infer the Paleobiology of Fossil Sharks

There are six methods commonly employed to reconstruct the paleobiology of fossil sharks.

1. The "classical" method of inferring the diet based on the teeth, as mentioned above. More discoveries have made it possible to reconstruct complete dentitions and infer the diet with greater accuracy. Complete dentitions, also called tooth sets, are a more solid framework with which to reconstruct the diets of the sharks than isolated teeth [14]. There are three types of tooth sets [14]. (a) In a natural tooth set, the jaw is preserved, and all of the teeth are in their original positions. This the best but also the rarest condition. (b) An associated tooth set is one based on the teeth of an individual shark, where the teeth are found displaced from their natural positions. This is also

rare and mostly associated with skeletal remains [10]. (c) An artificial tooth set can be constructed from a number of tooth types from one locality that are believed to belong to one species. The teeth probably come from different individuals. This is the main type of reconstruction.

2. The rare discovery of preserved articulated or disarticulated skeletons or parts thereof, including body proportions, gastric contents, and data on reproductive biology [11].
3. Bite marks on fossil bones or shark teeth embedded next to the fossilized skeletal remains of prey animals can also be used to provide direct evidence of predation or scavenging [15,16].
4. Stable isotopes can be used to reconstruct trophic positions [17,18].
5. Two-dimensional or three-dimensional computer modeling based on vertebral centra and morphometric comparisons with Recent sharks can provide information on body size and tooth shape [19,20].
6. The shape and morphology of the placoid scales can be used to reconstruct swimming abilities [21].

3. Materials and Methods

For this review, the literature was searched for information concerning the ecology and paleobiology of extinct Neogene shark species, as well as for the referral of fossil remains to extant species [22]. Although the focus of this paper is on Neogene shark species, when the first occurrence of extant species predates the Neogene, this is nevertheless also noted. Throughout this review, extinct shark species are labeled with a dagger symbol (†) for clarity. Geologic ages can be found in Figure 1.

Pleistocene		
Plioc.	Late	Piacenzian
	Early	Zanclean
Miocene	Late	Messinian
		Tortonian
	Middle	Serravallian
		Langhain
	Early	Burdigalian
		Aquitanian
Oligocene		
Eocene		

Figure 1. Stratigraphic table.

The fossil record of Recent species is documented. In addition, when remarkable information concerning the biology of Recent species has been discovered from fossil sources, e.g., a dietary shift, this is mentioned in the text. Otherwise, the reader is referred to the according literature, because details of the ecology of extant sharks are well documented elsewhere. Likewise, extant species are not considered because photos of them can be found in nearly every scientific or non-scientific book on sharks.

For each Neogene shark species, one fossil tooth has been illustrated, or, in the case of the extinct basking shark *Keasius parvus*, a gill raker (Figures 2 and 3). The latter species first appears in the Oligocene (Paleogene) and the identified raker is from this epoch simply because it was the best-preserved one available to the authors. However, a complete preserved tooth was not available for every taxon. The extinct Neogene shark species and the according methods used to infer their paleobiology are summarized in Table 1. Many of the teeth of the extinct Neogene shark species mentioned in this paper have a nearly global distribution. It was therefore decided not to list all of their fossil discovery localities, as was done for the fossil records of extant sharks, in order to constrain the length of this paper. Despite the large volume of research on fossil sharks undertaken during the past few decades, there are unresolved questions and different opinions, especially concerning the genus-level membership of some taxa. However, a discussion of the problems regarding Neogene taxa is beyond the scope of this paper, and it is not relevant to this review. Details of these debates can be found in the cited literature.

Table 1. Extinct (†) Neogene shark species and the methods used to infer their paleobiology.

Extinct Neogene Shark Species	Methods Used for Paleobiological Reconstruction
†*Mitsukurina lineata* (Probst)	Isotopes (δ^{66}Zn values)
†*Araloselachus cuspidatus* (Agassiz)	Isotopes (δ^{66}Zn values), skeletal remains
†*Carcharoides catticus* (Philippi)	Artificial tooth set
†*Carcharias gustrowensis* (Winkler)	Skeletal remains
†*Carcharodon hastalis* (Agassiz)	Bite traces on fossil dolphin skeleton, tooth height and width, skeletal remains, stomach content, isotopes (δ^{66}Zn and $\delta^{15}N_{EB}$ values)
†*Carcharodon hubbelli* Ehret, MacFadden, Jones, DeVries, Foster and Salas-Gismond	Vertebral centra, tooth height
†*Megalolamna paradoxodon* Shimada, Chandler, Lam, Tanaka & Ward	Tooth height and shape, paleoenvironment
†*Otodus* (*Megaselachus*) *megalodon* (Agassiz) and†*O.* (*M.*) *chubutensis* (Ameghino)	2D and 3D reconstructions, isotopes (δ^{66}Zn, $\delta^{18}O_p$, and $\delta^{15}N_{EB}$ values), vertebral centra, tooth height and width, plaeoenvironment, comparison with the extant great white shark (*Carcharodon carcharias*), placoid scales and tessellated calcified cartilage remains, marine mammal bones with bite traces from †*Otodus* teeth
†*Parotodus benedeni* (Le Hon)	Artificial tooth set, tooth shape and height, comparison with members of Lamnidae and Otodontidae
†*Keasius parvus* (Leriche)	Shape of gill rakers, skeletal remains
†*Megachasma applegatei* Shimada, Welton and Long	Tooth shape (including a landmark-based geometric morphometric analysis), paleonenvironment
†*Pachyscyllium distans* (Probst) and †*Pachyscyllium dachiardii* (Lawley)	Paleoenvironment
†*Hemipristris serra* Agassiz	Tooth size, artificial tooth set, paleoenvironment, isotopes (δ^{66}Zn value)
†*Galeocerdo aduncus* (Agassiz)	Preserved jaw fragment, bite marks on a †*Metaxytherium* carcass and on a crocodilian coprolite, isotopes (δ^{66}Zn value)
†*Physogaleus contortus* Gibbes	Tooth shape, teeth association with a cetacean carcass
†*Carcharhinus dicelmai* Collareta, Kindlimann, Baglioni, Landini, Sarti, Altamirano, Urbina & Bianucci	Tooth size, paleoenvironment

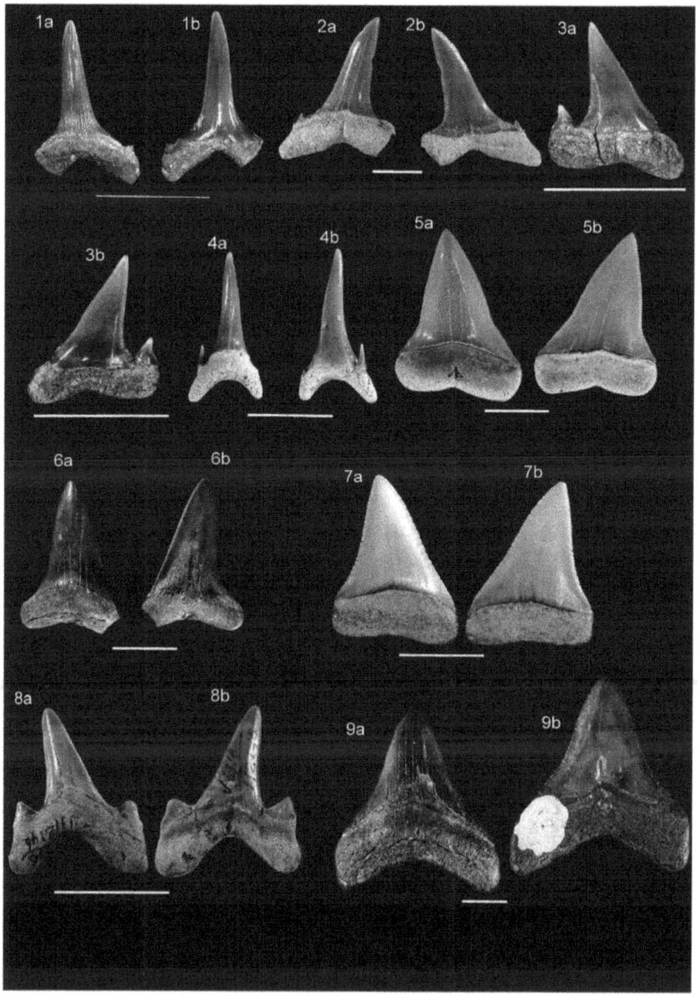

Figure 2. (**1a**,**1b**) *Mitsukurina lineata* (Probst). SMNS 97016/10, Miocene, Rengetsweiler, Baden-Württemberg, Germany. (**1a**) Lingual view; (**1b**) labial view. Scale: 10 mm. (**2a**,**2b**) *Araloselachus cuspidatus* (Agassiz). SMNS 97269, Miocene, Kühnring, Lower Austria. (**2a**) Lingual view; (**2b**) labial view. Scale: 10 mm. (**3a**,**3b**) *Carcharoides catticus* (Philippi). SMNS 97015/42, Miocene, Rengetsweiler, Baden-Württemberg, Germany. (**3a**) Lingual view; (**3b**) labial view. Scale: 10 mm. (**4a**,**4b**) *Carcharias gustrowenis* (Winkler). SMNS 97015/55, Miocene, Rengetsweiler, Baden-Württemberg, Germany. (**4a**) Lingual view; (**4b**) labial view. Scale: 10 mm. (**5a**,**5b**) *Carcharodon hastalis* (Agassiz). Broad-toothed morphotype. SMNS 97270, Miocene, Atacama Desert, Chile. (**5a**) Lingual view; (**5b**) labial view. Scale: 20 mm. (**6a**,**6b**) *Carcharodon hastalis* (Agassiz). "Narrow-toothed" morphotype. SMNS 55505, Miocene, Baltringen, Baden-Württemberg, Germany. (**6a**) Lingual view; (**6b**) labial view. Scale: 20 mm. (**7a**,**7b**) *Carcharodon hubbelli* Ehret, MacFadden, Jones, DeVries, Foster and Salas-Gismond. SMNS 97271, Miocene, Peru. (**7a**) Lingual view; (**7b**) labial view. Scale: 20 mm. (**8a**,**8b**) *Megalolamna paradoxodon* Shimada, Chandler, Lam, Tanaka & Ward. UCMP 112146, Miocene, Jewett Sand, Kern County, California, USA. (**8a**) Lingual view; (**8b**) labial view. Scale: 20 mm. Images courtesy of K. Shimada, used with permission. (**9a**,**9b**) *Otodus* (*Megaselachus*) *megalodon* (Agassiz). SMNS 97266, Miocene, Malta. (**9a**) Lingual view; (**9b**) labial view. Scale: 20 mm.

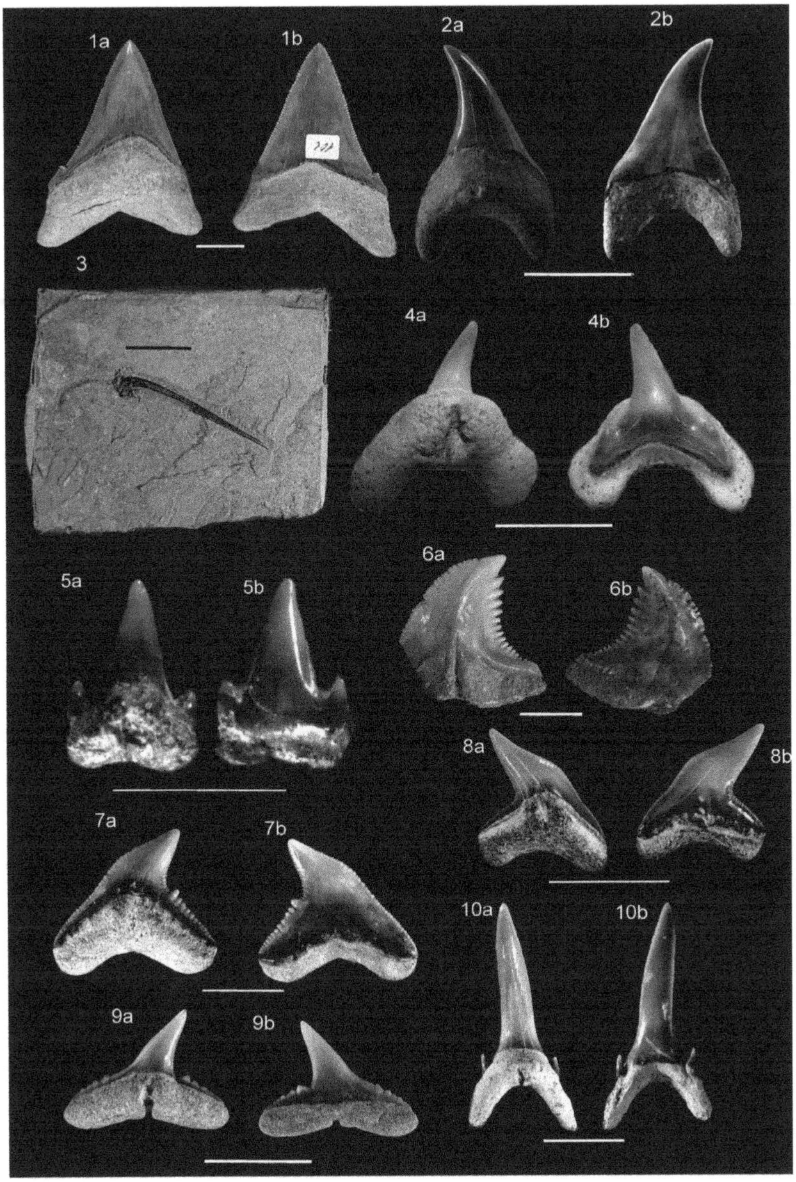

Figure 3. (**1a**,**1b**) *Otodus* (*Megaselachus*) *chubutensis* (Ameghino). SMNS 97267, Miocene, Lake Constance, Germany. (**1a**) Lingual view; (**1b**) labial view. Scale: 20 mm. (**2a**,**2b**) *Parotodus benedenii* (Le Hon). Miocene, Rengetsweiler, Baden-Württemberg, Germany. Specimen housed in a private collection. (**2a**) Lingual view; (**2b**) labial view. Scale: 20 mm. (**3**) *Keasius parvus* (Leriche). SMNS 80740/16, gill raker from the Bodenheim Formation, Oligocene. Rauenberg, Baden-Württemberg, Germany. Scale: 20 mm. (**4a**,**4b**) *Megachasma applegatei* Shimada, Welton and Long). LACM 122190, Miocene, Pyramid Hill Sand Quarry in southeastern San Joaquin Valley, California. Photos courtesy of Kenshu Shimada, used with permission. (**4a**) Lingual view; (**4b**) labial view. Scale: 5 mm. (**5a**,**5b**) *Pachyscyllium dachiardii* (Lawley). SMNS 56753, Miocene, Ursendorf, Baden-Württemberg, Germany. (**5a**) Lingual view; (**5b**) labial view. Scale: 5 mm. (**6a**,**6b**) *Hemipristris serra* Agassiz. SMNS

85944/1, Miocene, Baltringen, Baden-Württemberg, Germany. (**6a**) Lingual view; (**6b**) labial view. Scale: 10 mm. (**7a**,**7b**) *Galeocerdo aduncus* (Agassiz). SMNS 97268, Miocene, Rammingen, Baden-Württemberg, Germany. (**7a**) Lingual view; (**7b**) labial view. Scale: 10 mm. *Physogaleus contortus* (Gibbes). SMNS 97272, Miocene, Will Beach, Maryland, USA. (**8a**) Lingual view; (**8b**) labial view. Scale: 15 mm. (**9a**,**9b**) *Carcharhinus dicelmai* Collareta, Kindlimann, Baglioni, Landini, Sarti, Altamirano, Urbina & Bianucci. MUSM 4697, Miocene, Peru. (**9a**) Lingual view; (**9b**) labial view. Scale: 5 mm. Photos courtesy of Alberto Collareta, used with permission. (**10a**,**10b**) *Carcharias gustrowenis* (Winkler,). SMNS 97015/55, Miocene, Rengetsweiler, Baden-Württemberg, Germany. (**10a**) Lingual view; (**10b**) labial view. Scale: 10 mm.

4. Results

4.1. Paleobiology of Extinct Neogene Shark Species

Lamniformes Berg

Mitsukurinidae Jordan

†*Mitsukurina lineata* (Probst) (Figure 2(1a,1b))

This is possibly the ancestor to the Recent *M. owstoni* (Jordan). Teeth of †*M. lineata* are found in bathyal and neritic deposits from the Early and Middle Miocene of Europe and South Korea [3,23]. Δ^{66}Zn values for teeth from the Early Miocene of Baden-Württemberg, Germany, show a lifestyle similar to that of *Pseudocarcharias kamoharai* (Matsubara) (syn. *P. rigida*) [18]. The latter species feeds on bony fishes, squid, and shrimp [24], which is also the case for the Recent *M. owstoni* [24]. Although *M. owstoni* is a mostly bathyal shark, rarely occurring in shallow waters close to shore [24], the teeth of †*M. lineata* have also been found in neritic deposits, as mentioned above. The species possibly visited shallower waters in search of food or followed schools of fishes [25].

Odontaspididae Müller & Henle

†*Araloselachus cuspidatus* (Agassiz) (Figure 2(2a,2b))

There are differing opinions as to whether this species belongs to the genus *Carcharias* (see the extant *Carcharias taurus*) or to the extinct genus †*Araloselachus* [3,11,26]. Likewise, its relationship with the species †*Araloselachus vorax* (Le Hon), which had similarly shaped teeth, is not yet resolved [3,16,26–28]. †*Araloselachus cuspidatus* is known from Miocene neritic deposits in Europe, North America, and Central Asia [3], as well as from older deposits of Oligocene age [11,29]. Its teeth are very abundant. They have a grasping, odontaspid shape but with a broader crown and often larger size than in †*Carcharias contortidens* or *C. taurus*. Δ^{66}Zn values indicate that †*A. cuspidatus* was likely a higher-trophic-level piscivore than †*M. lineata* and *Pseudocarcharias kamoharai* (syn. *P. rigida*) [18], also supported by the larger tooth size of †*A. cuspidatus* [18]. A partial skeleton of †*A. cuspidatus*, including fetuses, from the Oligocene of Germany was illustrated and described [11]. An estimated body length of ca. 5 m was mentionedfor this specimen [11]. Adelophagy (intrauterine cannibalism), which is characterized by larger pups preying on smaller ones, is well documented among unborn pups of the extant *Carcharias taurus* [11,24,30]. This might also have occurred in †*A. cuspidatus* and could explain the large number of incomplete embryos recovered [11].

†*Carcharoides catticus* (Philippi) (Figure 2(3a,3b)).

Two species of †*Carcharoides* are known from the Neogene. Using both of these species, an artificial tooth set was constructed for †*Carcharoides catticus* (Philippi) [31]. Based on the tooth morphology of †*C. catticus*, this species was considered to be a synonym of *Triaenodon obesus* (Rüppell) [32]; however, other authors dealing with this species do not share this opinion and cite this species as †*C. catticus* [28,31]. Dried jaws or teeth of *T. obesus* were not available to the authors for comparison; therefore, the fossil teeth are treated here as †*C. catticus*. The reconstruction of the dentition shows similarities to the dentition of *Carcharias* and *Odontaspis* [31]; therefore, a piscivorous diet can be also assumed for †*C. catticus*. Weak ontogenetic heterodonty for members of †*Carcharoides* was mentioned [31]. The species †*C. catticus* first appears in the Oligocene. and teeth of up to Middle Miocene age can be found in the neritic sediments of Europe and North America [31,32].

Carchariidae Müller & Henle, 1838

†*Carcharias gustrowensis* (Winkler) (Figure 2(4a,4b)).

This species existed from the Oligocene to at least the Lower Miocene [10,28]. Hovestadt & Hovestadt-Euler (2010) [10] A partial skeleton of a gravid shark with eight fetuses, also associated with a myliobatoid tail spine and a chimaeroid dorsal fin spine was described from the Oligocene of Baden-Württemberg, Germany [10]. The variation in the length of the fin radials in †*C. gustrowensis* resembles the pectoral fin skeleton of *Carcharias taurus* [10]. The myliobatoid and chimaeroid spines are likely remains of prey that have pierced the skin or cartilage of the jaw area. In addition to the Oligocene and Miocene of Germany, the species is also known from the Miocene of the Netherlands, Belgium, and Hungary, as well as the USA (North Carolina and Chesapeake Bay) [28,33].

Lamnidae Müller & Henle, 1838

The *Carcharodon* complex

The most recent systematic arrangement of the tooth shape indicates that †*Carcharodon hastalis* (Early Miocene–Pleistocene) is the oldest member of this genus, followed by †*Carcharodon hubbelli* (Late Miocene) and lastly the extant species *Carcharodon carcharias* (Early Pliocene–Recent) [8,9].

†*Carcharodon hastalis* (Agassiz) (Figure 2(5a,5b,6a,6b))

Teeth of this species are common worldwide in temperate to tropical neritic deposits of Early Miocene to Pleistocene age [3,34]. The generic referral of this species remains debated. There are also some uncertainties at the species level, with a narrow-toothed morphotype as well as a broad-toothed one. There is therefore a discussion of whether two other broad-toothed species (†*C. plicatilis* and †*C. xiphodon*) (Figure 2(5a,5b)) are distinct from the narrow †*C. hastalis* tooth morphotype (Figure 2(6a,6b)) [9,16,25,32,35,36]. This morphological difference could represent sexual dimorphism or ontogenetic change [9]. Assuming that all the referred teeth belong to only one species, the maximum body size would have been 6–7.6 m, with anterior teeth up to 8.1 cm in height [32]. A partially complete, articulated skeleton of a †*C. hastalis* juvenile, including stomach contents, was documented from the Late Miocene of Peru [37]. The total body length of the immature specimen was estimated to be 2.3–2.4 m. The Meckel's cartilages are very similar to those of various extant Lamniformes (including *Carcharodon carcharias* and *Isurus* spp.). The teeth are distinctly more slender than the adult teeth of †*C. hastalis*, in agreement with the pronounced ontogenetic heterodonty recognized in this species) [37]. The stomach contents consisted of fishes, including the pilchard *Sardinops* cf. *sagax*. It is possible that individuals with the narrow-toothed morphology had a piscivorous lifestyle, whereas those with the broad-toothed morphotype had a diet primarily consisting of small marine mammals [37]. In the Pisco Formation, sixteen teeth of †*C. hastalis* were found in close contact with a balaenopterid whale skeleton [38]. A tooth of †*C. hastalis*, early in its development, which was completely penetrated by a myliobatiform caudal spine, could be verified from the Calvert Cliffs (USA) (8–18 Ma, Miocene) [16]. Bite traces were found on a well-preserved fossil dolphin skeleton from the Pliocene of Italy [39]. Most bite traces were caused by a shark with unserrated teeth and about 4 m in length and were attributed to †*C. hastalis* based on their morphology and the distribution of traces on the skeleton. Additionally, bite traces attributed to †*C. hastalis* on cetacean skeletons from the Zanclean (Early Pliocene) of South Africa were described [15]. In contrast to the bite trace record, $\delta^{15}N_{EB}$ values in Miocene-aged †*C. hastalis* teeth were similar to those of Pliocene and extant *C. carcharias*, but lower, more piscivore-like values were found in the Pliocene [17]. Congruently, $\delta^{66}Zn$ signals indicated that †*C. hastalis* teeth from the Early Miocene of Malta had a higher trophic position than teeth from the Early Pliocene of North Carolina. However, conspecific teeth from the Miocene of Baden-Württemberg, Germany, also indicated a lower trophic position, potentially indicating that the regional availability of different prey types influenced the diet [18]. The same result was recovered for †*Hemipristis serra* between the two Miocene localities, lending support to this hypothesis. However, another possible interpretation is that the previously mentioned tooth morphotypes were correlated with trophic signal. Based on material in collections, it seems that only the narrower morphotype was present in the Early Miocene of Baden-Württemberg [40]. Reasons underlying the extinction of †*C. hastalis* are unknown.

†*Carcharodon hubbelli* Ehret, MacFadden, Jones, DeVries, Foster, and Salas-Gismond (Figure 2(7a,7b)).

A well-preserved jaw containing 222 teeth and associated with a series of 45 vertebral centra was recovered from the Late Miocene Pisco Formation of Peru. The teeth show similarities to those of *C. carcharias* and †*C. hastalis*, and †*C. hubbelli* was therefore interpreted as an intermediate species between †*C. hastalis* and *C. carcharias* [8,9]. †*C. hubbelli* is also known form the Late Miocene of California, USA, and Chile [41,42].

The examination of the vertebral centra yielded an age of at least 20 years. Based on measurements of the teeth and vertebral centra, this specimen is estimated to have had a minimum total body length of 4.80–5.07 m. The growth of †*C. hubbelli* appears to have been slower than that of Recent great white sharks [8,9]. †*C. hubbelli* fed on marine mammals [9].

†Otodontidae Glickman, 1964.

†*Megalolamna paradoxodon* Shimada, Chandler, Lam, Tanaka, & Ward (Figure 2(8a,8b)).

This species is known from teeth from the Early Miocene of the USA (North Carolina, California), Japan, and Peru [43,44], as well as from Baden-Württemberg, Germany (as "*Lamna* sp.") [25]. All the deposits represent shallow-water shelf-type coastal environments [25,43,44]. The largest came from an individual measuring at least 3.7 m in total length [43]. Based on the shape of the anterior and lateral teeth, the diet of †*M. paradoxodon* may have included relatively large prey, such as medium-sized (ca. 0.5–1 m) fishes, captured with the anterior teeth and cut by the distal portion of the dentition to a size suitable for ingestion [43].

†*Otodus* (*Megaselachus*) *megalodon* (Agassiz) (Figure 2(9a,9b)) and †*O.* (*M.*) *chubutensis* (Ameghino) (Figure 3(1a,1b)).

In the past, these extinct species have been placed in various genera (*Carcharodon*, †*Procarcharodon*, †*Carcharocles*, †*Megaselachus*); they are currently placed in †*Otodus*, and †*Megaselachus* is considered to be a subgenus [3,16].

†*Otodus* is divided into two chronospiecies: †*O. (M.) chubutensis*, with lateral cusplets or only traces thereof, and †*O. (M.) megalodon*, which lacks lateral cusplets. In Early Miocene deposits, teeth with cusplets are more abundant than uncuspleted ones. Moving upwards through the Miocene profile, uncuspleted *Otodus* teeth increase in relative abundance and the cuspleted ones eventually disappear entirely [45] (pers. observ. O.H.). A definitive separation between all the teeth of the taxa †*O. chubutensis* and †*O. megalodon* is impossible, because a complex mosaic evolutionary continuum characterizes this transformation, particularly in the loss of the lateral cusplets [45]. The cuspleted and uncuspleted teeth of †*Otodus* (*Megaselachus*) spp. are therefore designated as chronomorphs, because there is broad overlap between them both morphologically and chronologically. The †*O. chubutensis*/*megalodon* problem was discussed in detail in the literature [16,33,45,46].

The large, triangular teeth of *Otodus* spp. are surely the most easily recognizable shark teeth. †*Otodus* teeth are found worldwide in neritic deposits of the Neogene Epoch (see Cappetta 2012) [40]. The teeth of †*O.* (*M.*) *chubutensis* can reach a height of 13 cm; the ones from †*O.* (*M.*) *megalodon* can reach 17 cm [33]. Based on the tooth size, the maximum body length of †*O.* (*M.*) *megalodon* was probably between 18 and 20 m [47]. Individuals of †*O* (*M.*)

megalodon were, on average, larger in cooler waters than those living in warmer waters [47]. In the shallow marine Miocene Gatún Formation of Panama, the majority of †*O. (M.) megalodon* teeth are very small [48]. The individuals from Gatún were mostly juveniles and neonates, with estimated body lengths of between 2 and 10.5 m. They therefore proposed that the Gatún Formation represents a paleo-nursery area for †*O.* (*M.*) *megalodon* [47]. Based on statistical analyses, the presence of five potential nurseries were noted, ranging from the Langhian (Middle Miocene) to the Zanclean (Pliocene) in age, with higher densities of individuals with estimated body lengths within the range typical of neonates and young juveniles [49]. However, it was argued by other authors that, while it is possible that neonatal †*O. (M.) megalodon* could have utilized nursery areas, the previously identified paleo-nurseries may reflect temperature-dependent trends rather than inferred life history strategies [46].

A viviparous reproductive strategy characterized by matrotrophy via oophagy is primitive for crown-lamniform sharks [50], resulting in large body sizes at birth. This is consistent with the inferred life history of †*O.* (*M.*) *megalodon* [51]. Incremental growth bands in the fossil vertebrae of a 9.2-m-long individual from the Miocene of Belgium (see below) reveal that the shark was born large at 2 m in length and that this specimen died at age 46 [50]. It was estimated that †*O.* (*M.*) *megalodon* had a lifespan of at least 88–100 years and that it had a slightly higher growth rate (19–23 cm/year) during the first 7 years of life relative to the remainder of its life (11–18 cm/year) [51]. Tessellated calcified cartilage remains beside the teeth of a ca. 11.7-m-long individual could be verified from the Miocene of Japan [21]. The morphology of each tessera (i.e., predominantly hexagonal) and the arrangement of tesserae as a tessellated calcified cartilage sheet in †*Otodus* (*M.*) *megalodon* are virtually identical to those of extant chondrichthyans [21]. Further, it was found that the size range of tesserae observed in the estimated 11.7-m-long individual of †*O.* (*M.*) *megalodon* is comparable to that of extant chondrichthyans, indicating that a larger body size does not necessarily produce larger tesserae [21]. This observation suggests that, in †*O.* (*M.*) *megalodon*, as in extant sharks, skeletal elements sheathed by tesserae developed through biomineralization along the margins of existing tesserae to form new tesserae, despite its gigantic body size [21]. The first reconstruction of the skeletal anatomy of †*Otodus* was performed in the year 1996 [52]. The most recent anatomical reconstructions were developed in 2020 and 2022 [19,20]. In 2020 a two-dimensional reconstruction of †*O. megalodon*, was produced, based on comparisons with extant Lamniformes [19]. The results suggest that a 16 m †*O.* (*M.*) *megalodon* likely had a head ~4.65 m long, a dorsal fin ~1.62 m tall, and a tail ~3.85 m high [19]. In 2022, a three-dimensional model of †*O. megalodon* was published [20]. The basis for the model was a vertebral column with 141 centra, belonging to a single, 9.2-m-long individual housed in the Royal Belgian Institute of Natural Sciences in Brussels, Belgium, in addition to comparisons with the skeleton of the Recent great white shark, *Carcharodon carcharias* [20]. This vertebral column was recovered from the Antwerp Basin in the 1860s; however, neither the locality nor an age has been specified for the specimen beyond a Miocene range (23 to 5.3 Ma) [20]. The reconstruction yielded a total length of 15.9 m and a body mass of 61,560 kg. The gape size was determined at different angles: the gape was 1.2 m in height at a 35° angle and 1.8 m in height at 75° angle. The gape was 1.7 m in width at both angles. The stomach volume was estimated to have been 9605 L. Prey up to 8 m in length could have been ingested whole, whereas larger prey (e.g., prey the size of the extant humpback whale, *Megaptera novaeangliae*) would have required additional processing [20]. It was calculated that the modeled †*O.* (*M.*) *megalodon* had an energy requirement of 98,175 kcal per day. Additionally, the mean absolute speed for the model was calculated at 1.4–4.1 m/s (=ca. 5.0–14.8 km/h) and the mean relative cruising speed was estimated at 0.09 body lengths per second [20]. However, other authors estimated lower cruising speeds for †*O. (M.) megalodon*, at 2.0 km/h, with a range of 0.9–3.0 km/h, based on the morphology of its placoid scales [21]. This authors also found that the general size of the placoid scales in the vast majority of extant pelagic lamniforms and carcharhiniforms, as well as in extinct lamniform taxa such as †*Cretoxyrhina*, †*Cretodus*, and †*Squalicorax*, was similar to the overall scale size of the much larger †*O. megalodon*. This demonstrates that the exceptionally large body sizes seen in †*O.* (*M.*) *megalodon* did not result in exceptionally large placoid scales [21]. Rather, new placoid scales of a similar small size were added throughout the ontogeny as the shark grew [21]. All the authors used the chronospecies name †*O. megalodon*, but there is no reason to assume that these data cannot be extrapolated to †*O. chubutensis* if of similar size.

†*Otodus* spp. were the top predators during the Miocene and Early Pliocene. There are many examples of marine mammal bones with bite traces from †*Otodus* teeth, including, e.g., small-sized mysticete cetaceans and pinnipeds from the Upper Miocene Pisco Formation (Southern Peru) [53] and a mysticete caudal vertebra from the Pliocene of North Carolina [54]. However, in the majority of cases, it remains unclear whether these feeding events on mammals document active hunting or scavenging [18]. Using enameloid-bound $\delta^{15}N$ ($\delta^{15}N_{EB}$) in †*Otodus* teeth, it could be determined that †*Otodus* (*M.*) *megalodon* as well as †*O.* (*M.*) *chubutensis* occupied a higher trophic level than that of any known marine species, extinct or extant [17]. The $\delta^{15}N_{EB}$ values show a large range for †*O. (M). megalodon*, which may reflect a generalist diet, with individuals feeding across many prey types and different trophic levels [17]. Many extant apex predatory sharks are also opportunistic in their prey selection [18]. Despite the bite traces on the mysticete bones noted above, the high $\delta^{15}N_{EB}$ values indicate that baleen whales were not the dominant prey of †*O. megalodon*, as extant baleen whales have a low trophic level and a correspondingly low δ^{15} N (Kast et al., 2022) [17]. $\delta^{66}Zn$ values derived from the tooth enameloid of †*O. megalodon* were used to find support for the previous conclusion that †*Otodus* spp. were apex predators feeding at a very high trophic level [18]. However, during the Early Pliocene, the †*Otodus* lineage represented by †*O.* (*M*). *megalodon* showed a considerable increase in mean $\delta^{66}Zn$ values in Atlantic populations, hinting at a reduced trophic position for the megatooth shark lineage in the Atlantic [18]. This could indicate a dietary shift, specifically that lower-trophic-level mammalian prey such as mysticetes (and perhaps herbivorous sirenians) may have become an important dietary component for Atlantic populations of †*O.* (*M*). *megalodon*. Extinct small- and medium-sized mysticetes (e.g., Cetotheriidae and various small balaenids and balaenopterids) were abundant during the Early Pliocene and were thus available as prey for *Otodus* spp. [18]). As can be seen, the isotopic results are partially in conflict with respect to the trophic level.

Thermophysiology is another important area of investigation concerning the paleobiology of the Neogene †*Otodus* spp. The question of endothermy in Neogene †*Otodus* sharks were examined using $\delta^{18}O_p$ values (P = phosphate) [55]. The measurements support endothermy in †*Otodus* (*M.*) *megalodon* and †*O.* (*M.*) *chubutensis* [55]. Based on their lower estimates of the cruising speed, it was suggested that the function of regional endothermy shifted from maintaining high cruising speeds to accelerating digestion and nutrient absorbtion during the evolution of gigantism in otodontids [21].

Regarding the extinction of †*Otodus* (*M.*) *megalodon*, two dates are reported in the newer literature: (1) before c. 2.6 Ma (Pliocene/Pleistocene boundary) [56]; (2) before c. 3.6 Ma (Early–Late Pliocene boundary) [57]. There are different opinions regarding the importance of competition with great white sharks or the extinction of small to mid-sized mysticete prey species, as possible drivers of the extinction [17,18,58]. Competition with hypercarnivorous odontocetes may have also played a role in the extinction process [18,57]. Concerning climatic changes as potential causes of the extinction, no evidence was found for direct effects of the global temperature [58]. It was noted that the gigantic body size, when combined with with the high metabolic cost of maintaining an elevated body temperature, may have made †*Otodus* species more vulnerable to extinction than the sympatric sharks that survived the Pliocene Epoch [55]. In summary, the reasons for the extinction of †*O.* (*M.*) *megalodon* are still unknown.

†*Parotodus benedenii* (Le Hon) (Figure 3(2a,2b)).

The teeth of †*Parotodus benedenii* can be up to 6 cm high. This species has been widely reported from the Early Oligocene through the Early Pliocene fossil beds of Europe (Belgium, Germany, Hungary, Italy, Malta, the Netherlands, Portugal, Slovakia, and Switzerland), Africa (Angola and South Africa), the Azores, and the United States, as well as from Australia, Japan, and New Zealand in the Western Pacific [16]. Despite its broad geographical distribution, this species is rare in Neogene deposits. During the Neogene, a clear increase in tooth size occurred, accompanied by a notable thickening of the root, which became very stout and globular [3]. Different authors illustrated an artificial tooth set of this species [32,33,59]. †*P. benedenii* was reconstructed as a large, hypercarnivorous shark that inhabited pelagic settings and fed primarily on large, soft prey and scavenged items [60]. Thus, some ecological partitioning did likely exist between †*P. benedenii* and other elasmobranch apex predators (including the extant species *Carcharodon carcharias*, *Carcharhinus leucas*, and *Galeocerdo cuvier* during the Pliocene) in Neogene mid-latitude seas. The body length of †*P. benedenii* was estimated at over 7 m [60] or between 6 and 7.5 m [32].

Cetorhinidae Gill

†*Keasius parvus* (Leriche) (Figure 3(3)).

This species was originally placed in the basking shark genus *Cetorhinus*. In 2013, the species was placed in the newly erected genus †*Keasius* [61], based on the shape of the gill rakers, the vertebral centra, and the dentition. †*K. parvus* existed from the Middle Eocene to Middle Miocene [62]. Remains have been found in Europe, Mexico, and Japan [61]. A partial skeleton of †*K. parvus* was described from the Oligocene (Rupelian) of Germany [62]. †*K. parvus* possessed a filter feeding apparatus similar to that of the extant *Cetorhinus maximus*, and it can be assumed that the species shared the same feeding habits. The aforementioned skeleton came from a ca. 2-m-long animal [62]. The maximum length of †*K. parvus* is estimated at 4.5–5 m [62].

Megachasmidae Taylor, Compagno & Struhsaker

†*Megachasma applegatei* Shimada, Welton and Long, 2014 (Figure 3(4a,4b)).

The teeth of this extinct megamouth shark are known from Late Oligocene–Early Miocene marine deposits of the Western USA [63]. †*M. applegatei* could have measured approximately 6 m in total length and likely had a broad diet, possibly including small fishes and planktonic invertebrates. The fossil record indicates that †*M. applegatei* either had a broad bathymetric tolerance or was a nektopelagic feeder over both deep- and shallow-water habitats [64]. An artificial tooth set of this species was examined via landmark-based geometric morphometric analysis [63]. The teeth were more variable in shape than those of the extant *Megachasma pelagios* (Taylor, Compagno, & Struhsaker). The teeth of the fossil species were probably arranged in the typical heterodont "lamnoid tooth pattern" [65] as in predatory lamniform sharks.

Carcharhiniformes Compagno

Scyliorhinidae Gill

†*Pachyscyllium distans* (Probst) and †*Pachyscyllium dachiardii* (Lawley) (Figure 3(5a,5b)).

Both catshark species lived contemporaneously and their teeth are widespread in the Miocene and Early Pliocene of Europe (e.g., Germany, Belgium, France, Netherlands, Portugal, Italy) [28,40,66]. Both species had very similar teeth; therefore, only a tooth from *P. dachiardii* was illustrated. The only known information regarding the paleoecology of these taxa is that both were thermophilic sharks) [28,66].

Hemigaleidae Hasse

†*Hemipristris serra* Agassiz, 1843 (Figure 3(6a,6b)).

The species is very widely distributed from the Late Oligocene (Chattian) through the Pleistocene in formations representing warmer-water regions of the Atlantic Ocean, Caribbean Sea, Mediterranean Sea, Indian Ocean, and Pacific Ocean [16]. An artificial tooth set for this species was published [32]. Whether †*H. serra* is the direct ancestor to the Recent *H. elongata* (Klunzinger) is questionable. Based on histological differences between its teeth and those of the extant *H. elongata* (Klunzinger, 1871), it was suggested that the generic reassignment of †*H. serra* is warranted [67]. †*H. serra* probably reached a length of c. 6 m [68], whereas the Recent species only attains lengths of 2.3–2.4 m [30]. There are some differences in tooth size through time and space. Teeth from the Early Miocene of Southern Germany have a maximum size of 31 mm in height and 25 mm in width [69], but teeth from the Early Pliocene of North Carolina, USA, reached a height of 41 mm and a width of 43 mm [32].

Based on the δ^{66}Zn composition, †*H. serra* from the Early Miocene of Malta occupied a higher trophic position than individuals from the Early Miocene of Baden-Württemberg, Germany. This is the same relative result recovered for individuals of †*Carcharodon hastalis* between the two localities; different prey availability or a shorter trophic chain in the German Molasse Basin may also be driving the observed pattern in this case. The Maltese specimens have a similar trophic position to †*Galeocerdo aduncus* [18].

Galeocerdonidae Poey

†*Galeocerdo aduncus* (Agassiz) (Figure 3(7a,7b)).

This ancient tiger shark is found worldwide in neritic sediments of Oligocene to Late Miocene age [70]. A preserved jaw fragment from the Miocene (8 to 18 Ma) of Calvert Cliffs, USA was illustrated [16]. The teeth are similar to those of the extant tiger shark *G. cuvier*, apart from differences concerning the serration as well as the size [70]. †*G. aduncus* teeth are smaller. However, some authors [32] placed this species in synonymy with the extant *G. cuvier* on the basis of similarities in overall morphology.

Fossil evidence from the Middle Miocene of the Styrian Basin (Austria) shows that †*G. aduncus* fed on a sirenian carcass (†*Metaxytherium* sp.) [71]. Other authors were also able to match tooth marks on a crocodilian coprolite to this species [72]. Zinc isotope values in the *Galeocerdo* lineage show no statistical variability with either age or locality, suggesting that tiger sharks occupied a similar trophic level and ecological role in the marine ecosystem since at least the Early Miocene [18]. †*G. aduncus* likely had a similar lifestyle to that of the extant *G. cuvier*, despite having smaller teeth.

†*Physogaleus contortus* (Gibbes) (Figure 3(8a,8b)).

Teeth are known from the Early and Middle Miocene of the Eastern United States (Maryland, North Carolina, and Virginia), Cuba, Panama, Peru, Germany, and Hungary [16]. The paleobiology of †*P. contortus* is largely unknown, although the slender, twisted tooth crowns are consistent with a largely piscivorous diet [16]. A sperm whale skeleton from the lower Calvert Formation of Popes Creek, Maryland, USA (Early to Middle Miocene) was associated with 37 †*P. contortus* teeth [16]. Although the teeth are exceptionally large, these sharks were far too small to have attacked and killed such substantial prey. Typically, such an association of teeth would be attributed to scavenging, although this is difficult to confirm. Based on the tooth morphology, it seems equally plausible that this tooth concentration represents †*Physogaleus* preying on small scavenging fishes attracted by the carcass [16].

Carcharinidae Jordan & Evermann

†*Carcharhinus dicelmai* Collareta, Kindlimann, Baglioni, Landini, Sarti, Altamirano, Urbina, & Bianucci (Figure 3(9a,9b)).

This newly described species is known from the Lower Miocene Chilcatay Formation of Peru (type locality) and from the Lower- to mid-Miocene (Burdigalian to Lower Langhian) Cantaure Formation of Venezuela. The latter locality suggests a trans-Panamanian distribution for this ancient species [73]. Given the dimensions of its teeth, †*C. dicelmai* was likely a diminutive carcharhinid and may have relied on small prey items (including, e.g., small bony fishes and invertebrates) that were individually captured and ingested through feeding actions that involved clutching [73]. †*C. dicelmai* may also have been an essentially thermophilic and very littoral shark [73].

Additional comments regarding fossil *Carcharhinus*: In the Pliocene of Tuscany, Italy, a fossil cetacean rib pierced by a partial requiem shark tooth (*Carcharhinus* sp.) was found [74]. Evidence for *Carcharhinus* sharks (mostly broad-toothed members of the genus) foraging upon cetaceans is preserved in the Mediterranean Pliocene fossil record in the form of bite traces and teeth associated with bones [74]. Species-level identifications were not provided.

4.2. The Fossil Records of Extant Shark Species

Hexanchiformes de Buen

Hexanchidae Gray

Hexanchus griseus (Bonnaterre)

Fossils of very large *Hexanchus* teeth (at least 25 mm in width) have been widely, if rarely, collected from Early Miocene to Pliocene sediments in Belgium, Chile, Italy, Japan, Malta, Peru, Portugal, and Spain, as well as California and North Carolina in the USA [16]. These were named as †*Hexanchus gigas* (Sismonda) by Kent [16] or as *Hexanchus* sp. by Purdy et al. [32]. Apart from the large size, they are similar to the teeth of the extant *H. griseus*. As yet, it is unclear whether they represent separate species or are conspecific.

A large *Hexanchus* tooth was associated with a cetacean skeleton (†*Cephalotropis coronatus* Cope) from the Late Miocene of Maryland, although it is uncertain whether this represents active predation or scavenging. Shark bite traces on a sirenian skeleton from Pliocene shoreface deposits of Tuscany (Italy) were mentioned, which can probably be attributed to an immature *H. griseus* [75].

Notorynchus cepedianus (Péron)

The fossil record of this extant species is not clear. Teeth of similar shape to those of *N. cepedianus* can be found from the Late Oligocene (Chattian) through the Late Miocene of Florida, Maryland, North Carolina, and Virginia, as well as Australia, Austria, the Azores, Belgium, Denmark, France, Germany, Japan, the Netherlands, Poland, Portugal, Slovakia, Spain, and Switzerland [16]. These fossil teeth have mostly been named as †*Notorynchus primigenius* (Agassiz) [40]. There are, however, differing opinions regarding whether †*N. primigenius* represents a distinct species [16] or is a synonym of *N. cepedianus* [32]. Interestingly, the geographic distribution of Recent *N. cepedianus* is quite unlike that of *Notorynchus* in the Neogene, with Recent members of this genus generally restricted to cool temperate waters, whereas, in the Neogene, the genus was also widely distributed in warm temperate and tropical waters [28].

Heptranchias perlo (Bonnaterre)

Fossil record: Early Miocene: Costa Rica [76]; Middle Miocene: Italy (Abruzzo, Parma) [77,78]; Late Miocene: Panama (Northern Panama) [79]; Portugal (Lisbon) (as "cf.") [80]; Late Miocene to Early Pliocene: Venezuela (Northeastern Venezuela) [81].

Squaliformes Goodrich

Centrophoridae Bleeker

Centrophorus granulosus (Bloch & Schneider)

Fossil record: Early to Middle Miocene: France (Vaucluse) [82]; Pliocene: Italy (Tuscany, Piedmont) and France (Le-Puget-sur-Argens) [83–85]. In the Miocene deposits of Europe and South America, many teeth have been named as *Centrophorus* cf. *granulosus* [86,87] since they show similarities to the extant *C. granulosus*. However, the dentition of the other 10 extant *Centrophorus* species is insufficiently known [22]. The assignment of isolated *Centrophorus* teeth to species is therefore not without problems.

Deania calcea (Lowe)

Fossil record: Early to Middle Miocene: France (Vaucluse) [82], Middle Miocene: Spain (Southeastern Spain) [88], Japan (Nagano Prefecture) (as "cf.") [89]; Early Pliocene: Italy (Parma) (as "cf.") [90].

Dalatiidae Gray
Dalatias licha (Bonnaterre)
Fossil record: Miocene: Italy (Sardinia) [91]; Early to Middle Miocene: France (Vaucluse, Southern France) [82,92,93], Colombia (Guajira Peninsula) (as "cf.") [94,95]; Middle Miocene: South Korea [23]; Early Miocene to Early Pliocene: Japan [96–99]; Late Miocene: Panama [79]; Pliocene: Japan [100]; Early Pliocene: France (Le-Puget-sur-Argens) [84]; Late Pliocene: Italy (Tuscany) [85].

Somniosidae Jordan
Scymnodon ringens du Bocage & Capello
Fossil record: Early Pliocene: Italy (Parma) [101]; Middle Pliocene: Italy (Romagna Apennines) (as "cf.") [102].

Somniosus rostratus (Risso)
Fossil record: Early Pliocene: Italy (Parma) [103].

Zameus squamulosus (Günther)
Fossil record: Early Pliocene: Italy (Parma) [101].

Orectolobiformes Applegate
Rhincodontidae Garman
Rhincodon typus Smith
Fossil record: Late Oligocene: USA (South Carolina) (as "cf.") [104]; Early Miocene: France (Occitania) (as *Rhincodon* sp.) [105]; Early to Middle Miocene: USA (Maryland, North Carolina) [32,106]; Late Miocene–Early Pliocene: Costa Rica [107].

Lamniformes Berg
Cetorhinidae Gill
Cetorhinus maximus (Gunnerus)
Fossil record: Following Hovestadt & Hovestadt-Euler [62], this extant species first appeared in the Middle Miocene, whereas Welton [108] cited the Late Miocene as the earliest occurrence. Material has been referred to this taxon from the Early to Middle Miocene: Japan (Saitama) [109]; Middle Miocene: Czech Republic (Kienberg) [110]; Late Miocene: USA (Oregon) (as "cf.") [108], USA (California) [111]; Late Miocene: Germany (Sylt, Lower Saxony [112,113]; Late Miocene to Early Pliocene: Chile (El Rincón) [114], Netherlands (Winterswijk-Almelo) [115]; Early Pliocene: Belgium (Kallo) [116], France (Le-Puget-sur-Argens, Anvers) [84,117]; Late Pliocene: Italy (Tuscany) [85].

Carchariidae Müller & Henle
Carcharias taurus Rafinesque
Teeth similar in shape to those of the extant *Carcharias taurus* (Rafinesque) can be found worldwide in Neogene neritic deposits. Teeth of this morphology are the most abundant shark teeth in these deposits and often occur en masse. Historically, Miocene teeth of this type have been identified as †*C. contortidens* (Figure 3(10a,10b)), but the relationship of this taxon with *C. taurus* is not completely clear [28]. Similar teeth from the Early Pliocene have been named as *C. taurus* [18,32]. One problem is that, despite their abundance, the teeth are often not completely preserved and therefore important details (e.g., lateral cusplets) are often missing.

Based on the δ^{66}Zn values, *Carcharias* teeth show a relatively stable trophic level and ecological niche through time and space [18], and a similar lifestyle to that of the extant

C. taurus can be assumed for the Miocene representatives, despite the controversial species-level classification. Details of the biology of *C. taurus* can be found in Ebert et al. [30]. Today, this species is distributed in nearly all warm and tropical waters apart from the Eastern and Central Pacific [30]. During the Miocene and part of the Pliocene, members of the genus *Carcharias* (probably *C. taurus*) also occupied waters off the western coast of South America, where, today, the species is absent [118]. The latter authors suggested that the local extinction of *Carcharias* was the consequence of a drop in global temperatures during the Middle Pliocene and Pleistocene, accompanied by a coeval drop in sea level that reduced the shelf area and therefore the suitable habitat for this species. The establishment of the Panamanian isthmus prevented the later migration of *C. taurus* from the north [118].

Odontaspididae Müller & Henle

Odontaspis ferox (Risso)

Fossil record: Early Miocene: Chile (Central Chile); Middle Miocene: USA (North Carolina) [32,119]; Middle Miocene–Pliocene: Chile (Northern Chile) [42]; Late Miocene–Early Pliocene: Venezuela [81]; Early Pliocene: USA (North Carolina) [32]; Late Pliocene: Italy (Tuscany) [85].

Pseudocarchariidae Taylor, Compagno & Struhsaker

Pseudocarcharias kamoharai (Matsubara)

Fossil record: Early Miocene: Germany (Baden-Württemberg, Bavaria) [40,120], Austria (Upper Austria) [120], Hungary [121], Switzerland (Schaffhausen) [122]; Middle Miocene: Italy (Parma) [123]; Late Miocene: Portugal (Alvalade) (as "cf.") [124]; Late Miocene–Early Pliocene: Venezuela [81].

Alopiidae Bonaparte

Alopias superciliosus Lowe

Fossil record: Oligocene: Germany (Bavaria) (as "cf.") [125]; Early Miocene: USA (North Carolina) [126], Peru [44], Colombia as "cf.") [94]; Early Miocene to Early Middle Miocene: Japan [96]; Middle Miocene: Netherlands [127]; Middle Miocene to Early Pliocene: USA (Florida) [128]; Late Miocene: Panama [79,129], Portugal (Alvalade Basin, Lisbon) (as "cf.") [130,131]; France (Luberon) (as "cf.") [93]; Late Miocene–Early Pliocene: Venezuela, Costa Rica [81,107]; Pliocene: Italy (Tuscany) [132].

Alopias vulpinus (Bonnaterre)

Fossil record: Miocene: Myanmar [133], India (Orissa) [134]; Early Miocene: Portugal (Algarve) [135]. There are also many occurrences of this taxon in the literature with "cf." or "aff." originating from deposits dating from the Oligocene [44,104,130,136]. The fossil record of *A. vulpinus* therefore requires reassessment.

Lamnidae Müller & Henle

Lamna nasus (Bonnaterre)

Fossil record: Late Miocene: Netherlands (Liessel) [137]; Early Pliocene: Belgium (Kallo) [116]; Late Pliocene Italy (Tuscany) [138].

Isurus oxyrinchus Rafinesque

This species is noted in sediments dating from the Oligocene [136]. It is known from many deposits in Germany, Belgium, France, Italy, Switzerland, USA, Japan, Chile, and Africa [3]. Fossil teeth similar in shape to the extant *I. oxyrinchus* have sometimes been named as †*Isurus desori* (Agassiz) [69]. At the moment, it is not clear if †*I. desori* is a valid species or is a synonym of *Isurus oxyrinchus*.

Isurus paucus Guitart-Manday

Fossil record: Early Miocene to Early Middle Miocene: Japan (Central Japan) [139]; Middle Miocene–Pliocene: possibly Chile (Northern Chile) [42].

Carcharodon carcharias (Linnaeus)

The extant great white shark first appeared in the Miocene or Early Pliocene [3,16]. For details on the biology of the extant *C. carcharias*, see Domeier [140]. The teeth occur worldwide in neritic sediments. In a few cases, predatory or scavenging behavior of fossil *C. carcharias* has been documented in the fossil record, and, as with observations on the extant *C. carcharias*, cetaceans were important prey species [15,16]. Cigala-Fulgosi [141] described the skeleton of an extinct dolphin with bite traces attributed to *C. carcharias* from the Pliocene of Italy (Piacenza). To date, there are no studies documenting piscivory in *C. carcharias* in the fossil record [16]. The δ^{66}Zn isotopic results indicate an increase in the trophic position of *C. carcharias* from the Early Pliocene to the Recent [18]. In a comparison between Recent and fossil *Carcharodon carcharias*, both mysticete and odontocete cetaceans appear to have been equally represented in the diet of this species during the Pliocene. In contrast, extant great white sharks primarily attack small odontocetes and only rarely attack mysticetes. This change could be due to both the general reduction in the body size of great white sharks over time, as well as the diminished diversity of the cetacean assemblage [142]. A sample of fossil teeth from Spain indicates that large *C. carcharias* close to 7 m long or larger were relatively common in the Early Pliocene [143]. Villafaña et al. [144] described a paleo-nursery area of the great white shark from the Pliocene of Chile. Fossil teeth of *C. carcharias* can often be found in the same deposits as the extinct megatooth shark *Otodus* (*Megalselachus*) *megalodon*—for example, in the Late Miocene/Early Pliocene of Chile [114]. This suggests that both sharks co-existed [143]. However, no direct interaction or competition between these two apex predators has been documented.

Carcharhiniformes Compagno

Triakidae Gray

Galeorhinus galeus (Linnaeus)

Fossil record: Late Eocene: USA (North Carolina) [145]; Early Miocene: USA (North Carolina) [126]; Late Miocene: Panama (as "cf.") [79]; Late Miocene–Early Pliocene: Chile (Bahía Inglesa) [114]; Early Pliocene: South Australia (as *Galeorhinus* cf. *australis*) [146]; Late? Pliocene: USA (California) (as *Galeorhinus zyopterus*) [147]; Late Pliocene: Chile (Valparaíso) [148].

Galeocerdonidae Poey

Galeocerdo cuvier (Péron & Lesueur)

Fossil record: Early Miocene: India (Gujarat) [149]; Middle Miocene: Hungary (Nyirád) [150], USA (Florida) [70]; Middle Miocene–Middle Pliocene: [151]; Late Miocene: Panama (Lago Bayano) [129]; Late Middle to Early Late Miocene: Panama (Central Panama) [152]; Late Miocene: Borneo (Brunei Darussalam) [153]; Pliocene: USA (Florida, North Carolina) (Webb & Tessmann 1968; Maisch et al., 2018) [154,155], Angola [156]; Early Pliocene: Libya [157]; Late Early/Early Late Pliocene: Italy (Tuscany) [158].

Carcharinidae Jordan & Evermann

Carcharhinus amblyrhynchoides (Whitley)

Fossil record: Late Miocene: Borneo (Brunei Darussalam) [159].

Carcharhinus amblyrhynchos (Bleeker)

Fossil record: Late Miocene: Borneo (Brunei Darussalam) [159].

Carcharhinus albimarginatus (Rüppell)

Fossil record: Late Miocene–Early Pliocene: Chile (North Coast) [114], Ecuador (Camarones River) [160]; Middle Miocene–Pliocene: Chile (Northern Chile) [42]; Pliocene: Chile (Bahía Inglesa) (Long 1993) [114].

Carcharhinus amboinensis (Müller & Henle)
Fossil record: Late Miocene: Borneo (Brunei Darussalam) [159].

Carcharhinus brachyurus (Günther)
Remarks and fossil record: This species can be traced back to the Early Miocene [161]. The species were found in a lot of Neogene and Pleistocene localities in Europe, North and South America, Australia, and Japan [161]. It had an Early Miocene East Pacific–Central West Atlantic center of origin [161]. The present-day distributional pattern of *C. brachyurus* is the product of historical biogeographic processes and likely reflects major changes in the global ocean system, including the closure of major seaways and the emergence of new oceanic circulation patterns [161]. Landini et al. [44,161,162] also identified the oldest copper shark nursery area in the East Pisco Basin of Peru, from the Early Miocene of the Chilcatay Formation and the Late Miocene of the Pisco Formation.

Carcharhinus brevipinna (Müller & Henle)
Fossil record: Miocene: India (Orissa) [134]; Late Miocene: Panama (Lago Bayano) [129]; Middle Miocene to Early Pliocene: USA (Florida) (as "cf.") [128].

Carcharhinus falciformis (Bibron in Müller & Henle)
Fossil record: Early to Late Miocene: Malta [67]; Middle Miocene: India (Kutch) [163], USA (North Carolina) [32]; Middle Miocene to Early Pliocene: USA (Florida) [128]; Late Miocene: Borneo (Brunei Darussalam) [159], Panama (Northern Panama, Lago Bayano) [129,164]; Late Miocene–Early Pliocene: Costa Rica [107]; Pliocene: USA (North Carolina) [155]; Early Pliocene: Italy (Tuscany) [165].

Carcharhinus glaucus (Linnaeus) (syn. *Prionace glauca*, see da Silva Rodrigues-Filho et al. [166].
Fossil record: Miocene: Sri Lanka [167]; Middle Miocene–Pliocene: Chile (Northern Chile) [42]; Late Miocene: ?Belgium (Antwerp International Airport) [168]; Late Miocene to Early Pliocene: Chile (Northern Chile) [169]; Early Pliocene: Italy (Parma) [90]; Late Pliocene: Italy (Umbria, Tuscany) [85,170].

Carcharhinus leucas (Valenciennes in Müller and Henle)
Fossil record: Early Miocene: Egypt (Moghra) [171], Peru (Zamaca) [44]; Middle Miocene: India (Kutch) [163], USA (North Carolina) [32]; Middle Miocene to Early Pliocene: USA (Florida) [128]; Middle Miocene–Middle Pliocene: Venezuela [151]; Late Miocene: Panama (Northern Panama) [164], Portugal (Alvalade Basin) (as "cf.") [124]; Late Miocene: Peru (Pisco Basin) [172]; Pliocene: Italy (Tuscany) [173], USA (Florida) [154]; Early Pliocene: USA (North Carolina [32]; Canary Islands (Gran Canaria, Fuerteventura) [174], South Africa (Langebaanweg) [175].

Carcharhinus limbatus (Müller & Henle)
Fossil record: Miocene: India (Orissa) [134]; Early Miocene: USA (Delaware) (Purdy 1998) [176]; Early Miocene to Late Pliocene: Colombia (Guajira Peninsula) (as "cf.") [95]; Middle Miocene to Early Pliocene: USA (Florida) [128]; Early Pliocene: Italy (Tuscany) [177].

Carcharhinus longimanus (Poey)
Fossil record: Early Miocene: India (Kathiawar, Piram Island, Orissa) [178,179]; Pliocene: Italy (Tuscany) [173], Spain (Alicante) [180]. Cappetta [181] identified a tooth from the Pliocene of North Carolina, USA, as *Pterolamiops longimanus*. *Pterolamiops* is a

junior synonym of *Carcharhinus* [182], but, according to Purdy et al. [32], Cappetta's tooth may belong to *C. leucas*.

Carcharhinus macloti (Müller and Henle)

Fossil record: Miocene: India (Orissa) [134]; Early Miocene: Brazil (Northeastern Amazonia) (as "cf.") [183], Peru (East Pisco Basin) [73]; Middle Miocene: USA (North Carolina) [32]; Late Miocene: Peru (Cerro Colorado) [184], Portugal (Lisbon) [80].

Carcharhinus obscurus (Lesueur)

Fossil record: Early Miocene: Egypt (Moghra) [171]; Mexico (Baja California) (as "cf.") [185]; Venezuela (as "cf.") [186]; Early to Middle Miocene: Cuba [187]; Middle Miocene: Grenada (Carriacou) [188]; Middle to Late Miocene: Ecuador (Carretera Flavio Alfaro) [160]; Middle Miocene–Middle Pliocene: Venezuela [151]; Middle Miocene–Pliocene: Chile (Northern Chile) [42]; Late Miocene: Portugal (Alvalade Basin) (as "cf.") [124], Panama (Northern Panama, Lago Bayano) [129,164]; Pliocene: Italy (Tuscany) [173]; Early Pliocene: USA (North Carolina) [32].

Carcharhinus perezi (Poey)

Fossil record: Early Miocene: Brazil (North Brazil) [189], USA (Delaware) [176]; Early to ?Middle Miocene: Venezuela (Falcón Basin) [190]; Early Miocene to Late Pliocene: Colombia (Guajira Peninsula) (as "cf.") [95]; Middle Miocene: USA (North Carolina) [32]; Early to Middle Miocene: Cuba [187]; Late Miocene: Panama (Northern Panama) [164], Portugal (Alvalade Basin) [124]; Pliocene: Italy (Tuscany) [173]; Early Pliocene: USA (North Carolina) [32].

Carcharhinus plumbeus (Nardo)

Fossil record: Early Miocene: Italy (Piedmont) [191]; Middle Miocene: USA (North Carolina) [32]; Middle Miocene to Early Pliocene: USA (Florida) [128]; Middle Miocene–Middle Pliocene: Venezuela [151]; Late Miocene: Panama [164], Portugal (Alvalade Basin) (as "cf.") [124]; Pliocene: Italy (Tuscany) [173]; Early Pliocene: USA (North Carolina) [32].

Carcharhinus sealei (Pietschmann)

Fossil record: Late Miocene: Borneo (Brunei Darussalam) [159].

Glyphis glyphis (Müller & Henle)

Fossil record: Early Miocene to Pliocene: Portugal [192]; Late Miocene: Borneo (Brunei Darussalam) (as "cf.") [159]; Pliocene: Italy (Toscana) [193].

Negaprion brevirostris (Poey)

Fossil record: Early Miocene: India (Orissa) [179], Peru (Zamaca) [44]; Early to Middle Miocene: Cuba [187]; Middle to Late Miocene: Ecuador [160]; Middle Miocene–Middle Pliocene: Venezuela [151]; Middle Miocene to Early Pliocene: USA (Florida) [128]; Late Miocene: Panama (Northern Panama, Lago Bayano) [129,164], Peru (Cerro Colorado) [184]; Pliocene: Angola (as "cf.") [156], USA (Florida, North Carolina) [154,155].

Sphyrnidae Gill

Sphyrna media (Linnaeus)

Fossil record: Early Miocene: Brazil (Northeastern Amazonia) (as "cf.") [183]; Middle Miocene: USA (North Carolina) (as "cf.") [32]; Late Miocene: Peru (Cerro Colorado) [184]; Pliocene: USA (North Carolina) (as "cf.") [32], Ecuador [160]; Late Pliocene–Pleistocene: Ecuador (Punta Canoa) [160].

Sphyrna mokarran (Rüppell)

Fossil record: Early Miocene: Cuba (Domo de Zaza) [194]; Middle Miocene to Early Pliocene: USA (Florida) [128]; Late Miocene: Panama (Lago Alajuela, Northern Panama, Lago Bayano) [129,152,164,195], Borneo (Brunei Darussalam) (as "cf.") [159].

Sphyrna zygaena (Linnaeus)

Teeth similar to this species have been found in sediments dating from the Early Miocene and younger [28]. However, there is debate as to whether these teeth belong to *S. zygaena* or to *Sphyrna laevissima* (Cope), described from the Miocene of Maryland, USA [28,32].

5. Outlook and Conclusions

Despite a fossil record consisting mostly of teeth, new findngs and methods have increased our knowledge of fossil shark species as well as the fossil records of extant species. In particular, isotopic analyses and computer-based 2D and 3D reconstructions are valuable tools for the study of fossil shark teeth. Paleobiological details surpassing descriptions of the teeth are known for a total of 19 extinct Neogene shark species, with most of the research focused on the well-known, large-bodied *Otodus megalodon*. Aside from the latter taxon, there are no hypotheses developed to date regarding potential causes underlying the extinction of these shark species; however, climate change and habitat loss have been suggested [196]. Concerning the fossil records of the more than 500 extant shark species, 38 species could be verified as present in the Neogene record. Four of these 38 species (*Alopias superciliosus, Galeorhinus galeus, Rhincodon typus,* and possibly *Alopias vulpinus,* 11%) first appeared during the Paleogene. For five extant species (*Carcharias taurus, Hexanchus griseus, Isurus oxyrinchus, Notorynchus cepedianus, Sphyrna zygaena*), the relationship of the extant and fossil forms is not clear. Figures 4 and 5 show the phylogenetic relationships and summarize the stratigraphic ranges of the species discussed in the text. The taxa are divided into Charchariniformes (Figure 5) and non-Carcharhiniformes (Figure 4) for readability. Determining the exact number of shark species present during the Neogene is highly speculative, if not impossible, although it can be assumed that the ancient diversity was similar to the extant diversity, with the addition of now extinct taxa. Reasons for this lack of knowledge include collecting bias (especially concerning minute-sized teeth), incomplete preservation of the teeth, and the poorly known dentition of extant relatives (here, also especially the small species with minute-sized teeth and also the presence or absence of different forms of heterodonty). Sometimes, only one tooth with a different shape is found in a sample, which is not enough for a reliable taxonomic diagnosis, for example, "*Carcharhinus* sp." from Äpfingen, Baden-Württemberg [197].

The implementation of the new methods mentioned herein, extensive collection (especially of minute teeth), and detailed descriptions of the dentition of Recent species will enhance our knowledge of shark evolution and the paleobiology of fossil sharks.

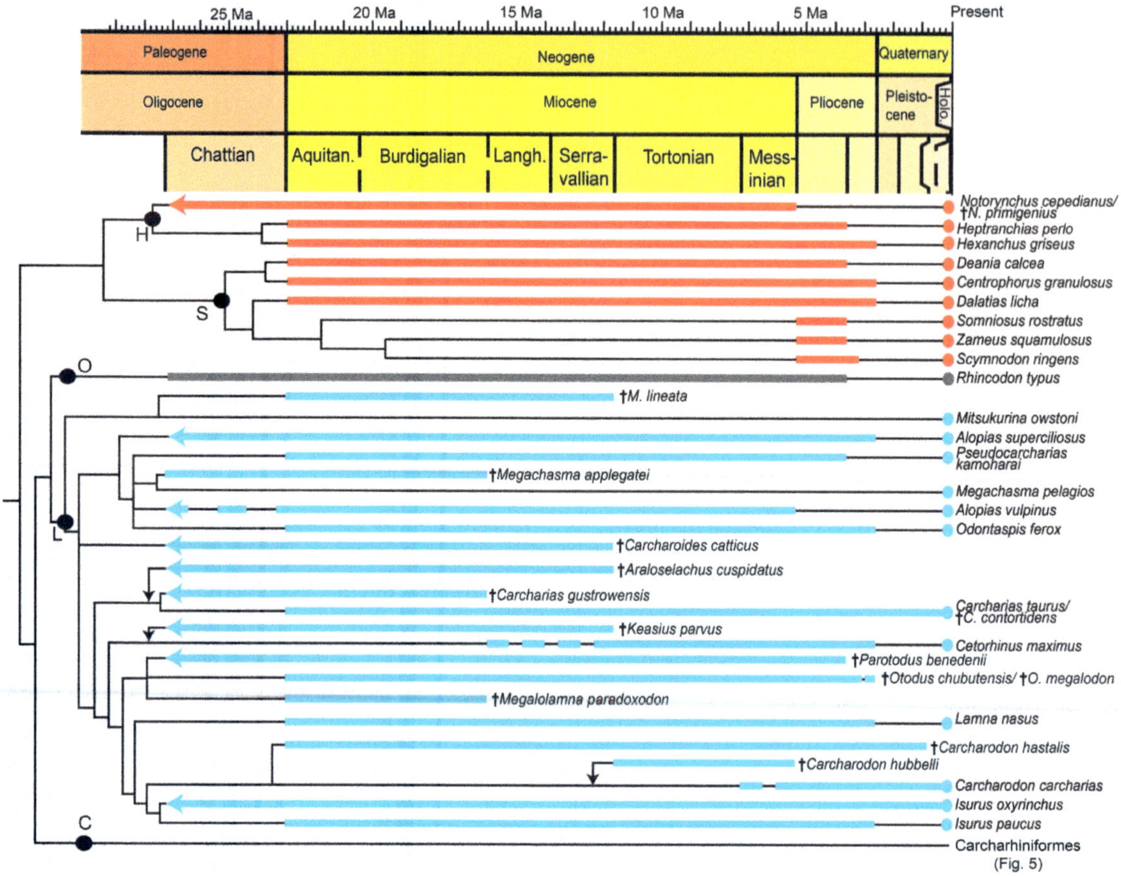

Figure 4. Relationships and stratigraphic ranges of non-carcharhiniform species discussed in the text. Topology derived from Stein et al. [198] for extant species, with position of extinct taxa following the review presented here. Branch arrows indicate phylogenetic uncertainty; range arrows indicate taxa that appeared prior to the Late Oligocene; and dashed range lines indicate stratigraphic or taxonomic uncertainty. Node positions not to scale. "Fig. 5" refers to Figure 5. C, Carcharhiniformes; H, Hexanchiformes; L, Lamniformes; O, Orectolobiformes; S, Squaliformes.

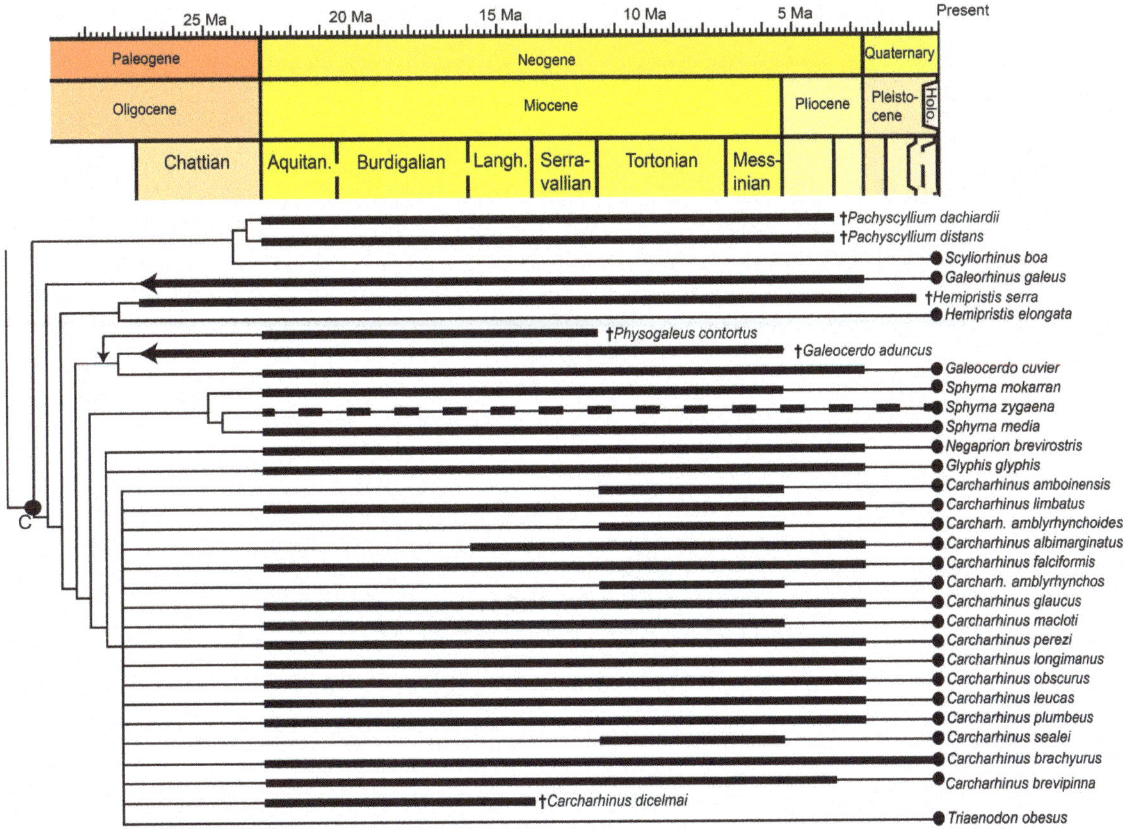

Figure 5. Relationships and stratigraphic ranges of carcharhiniform species discussed in the text. Topology derived from Stein et al. [198] for extant species, with position of extinct taxa following the review presented here. Branch arrows indicate phylogenetic uncertainty; range arrows indicate taxa that appeared prior to the Late Oligocene; and dashed range lines indicate stratigraphic or taxonomic uncertainty. Node positions not to scale. C, Carcharhiniformes.

Author Contributions: Conceptualization, O.H., E.E.M. and M.W.R.; methodology, O.H., E.E.M. and M.W.R.; investigation, O.H.; resources, O.H., E.E.M. and M.W.R.; data curation, O.H., E.E.M. and M.W.R.; writing—original draft preparation, O.H.; writing—review and editing, E.E.M.; visualization, E.E.M.; supervision, O.H., E.E.M. and M.W.R.; project administration, O.H., E.E.M. and M.W.R. All authors have read and agreed to the published version of the manuscript.

Funding: This research received no external funding.

Institutional Review Board Statement: Not applicable.

Data Availability Statement: No new data were created or analyzed in this study. Data sharing is not applicable to this article.

Acknowledgments: We wish to thank Alberto Collareta (University of Pisa, Italy) and Jürgen Pollerspöck (Stephansposching, Bavaria, Germany), as well as Kenshu Shimada (DePaul University, Ilinois, USA), for the permissions to use their photographs of shark teeth. We also thank two anonymous reviewers for their suggested improvements to the manuscript.

Conflicts of Interest: The authors declare no conflicts of interest.

Abbreviations

LACM	Natural History Museum of Los Angeles County, Los Angeles, California
MUSM	Museo de Historia Natural de la Universidad Nacional Mayor de San Marcos, Jesús María, Lima, Peru
SMNS	Staatliches Museum für Naturkunde Stuttgart, Stuttgart, Germany
UCMP	University of California at Berkeley, Museum of Paleontology Berkeley, California, USA

References

1. Andreev, P.S.; Coates, M.I.; Shelton, R.M.; Cooper, P.R.; Smith, M.P.; Sansom, I.J. Upper Ordovician chondrichthyan-like scales from North America. *Palaeontology* **2015**, *58*, 691–704. [CrossRef]
2. Benton, M.J. *Vertebrate Palaeontology*, 3rd ed.; Blackwell Publishing Company: Oxford, UK, 2005; pp. 164–169.
3. Cappetta, H. *Handbook of Paleoichthyology, Vol. 3E: Chondrichthyes · Mesozoic and Cenozoic Elasmobranchii: Teeth*; Verlag Dr. Friedrich Pfeil: Munich, Germany, 2012.
4. Compagno, L.J.V. Phyletic relationships of living sharks and rays. *Am. Zool.* **1977**, *17*, 303–322. [CrossRef]
5. Maisey, J.G. What is an 'elasmobranch'? The impact of palaeontology in understanding elasmobranch phylogeny and evolution. *J. Fish Biol.* **2012**, *80*, 918–951. [CrossRef]
6. Janvier, P.; Pradel, A. Elasmobranchs and their extinct relatives: Diversity, relationships, and adaptations through time. In *Fish Physiology (Vol. 34), Physiology of Elasmobranch Fishes: Structure and Interaction with Environment*; Shadwick, R.E., Farrell, A.P., Brauner, C.J., Eds.; Academic Press: Cambridge, MA, USA, 2015; pp. 1–17.
7. Maisey, J.G.; Janvier, P.; Pradel, A.; Denton, J.S.S.; Bronson, A.; Miller, R.; Burrow, C.J. Doliodus and pucapampellids: Contrasting perspectives on stem chondrichthyan morphology. In *Evolution and Development of Fishes*; Johanson, Z., Underwood, C., Richter, M., Eds.; Cambridge University Press: Cambridge, UK, 2019; pp. 87–109. [CrossRef]
8. Ehret, D.J.; Hubbell, G.; Macfadden, B.J. Exceptional preservation of the white shark *Carcharodon* (Lamniformes, Lamnidae) from the early Pliocene of Peru. *J. Vertebr. Paleontol.* **2009**, *29*, 1–13. [CrossRef]
9. Ehret, D.J.; Macfadden, B.J.; Jones, D.S.; Devries, T.J.; Foster, D.A.; Salas-Gismondi, R. Origin of the white shark *Carcharodon* (Lamniformes: Lamnidae) based on recalibration of the Upper Neogene Pisco Formation of Peru. *Palaeontology* **2012**, *55*, 1139–1153. [CrossRef]
10. Hovestadt, D.C.; Hovestadt-Euler, M. A partial skeleton of *Carcharias gustrowensis* (Winkler, 1875) (Chondrichthyes, Odontaspididae) including embryos, a chimaeriod dorsal fin spine and a myliobatoid tail spine from the Oligocene of Germany. *Cainozoic Res.* **2010**, *7*, 83–97.
11. Hovestadt, D.C. A partial skeleton of *Carcharias cuspidatus* including embryos from the Oligocene of Germany. *Cainozoic Res.* **2022**, *22*, 25–36.
12. Bazzi, M.; Campione, N.E.; Kear, B.P.; Pimiento, C.; Ahlberg, P.E. Feeding ecology has shaped the evolution of modern sharks. *Curr. Biol.* **2021**, *31*, 5138–5148. [CrossRef] [PubMed]
13. Cooper, J.A.; Griffin, J.N.; Kindlimann, R.; Pimiento, C. Are shark teeth proxies for functional traits? A framework to infer ecology from the fossil record. *J. Fish Biol.* **2023**, *103*, 798–814. [CrossRef] [PubMed]
14. Welton, B.J.; Farish, R.F. *The Collector's Guide to Fossil Sharks and Rays from the Cretaceous of Texas*; Before Time: Lewisville, TX, USA, 1993.
15. Govender, R. Shark-Cetacean trophic interaction, Duinefontein, Koeberg, (5 Ma), South Africa. *S. Afr. J. Sci.* **2015**, *111*, 178–184. [CrossRef]
16. Kent, B.W. The Cartilaginous Fishes (Chimaeras, Sharks, and Rays) of Calvert Cliffs, Maryland, USA. In *The Geology and Vertebrate Paleontology of Calvert Cliffs, Maryland*; Godfrey, S.J., Ed.; Smithsonian Institution Scholarly Press: Washington, DC, USA, 2018; pp. 45–157.
17. Kast, E.R.; Griffiths, M.L.; Kim, S.L.; Rao, Z.C.; Shimada, K.; Becker, M.A.; Maisch, H.M.; Eagle, R.A.; Clarke, C.A.; Neumann, A.N.; et al. Cenozoic megatooth sharks occupied extremely high trophic positions. *Sci. Adv.* **2022**, *8*, eabl6529. [CrossRef]
18. McCormack, J.; Griffiths, M.L.; Kim, S.L.; Shimada, K.; Karnes, M.; Maisch, H.; Pederzani, S.; Bourgon, N.; Jaouen, K.; Becker, M.A.; et al. Trophic position of Otodus megalodon and great white sharks through time revealed by zinc isotopes. *Nat. Commun.* **2022**, *13*, 2980. [CrossRef] [PubMed]
19. Cooper, J.A.; Pimiento, C.; Ferrón, H.G.; Benton, M.J. Body dimensions of the extinct giant shark *Otodus megalodon*: A 2D reconstruction. *Sci. Rep.* **2020**, *10*, 14596. [CrossRef] [PubMed]
20. Cooper, J.A.; Hutchinson, J.R.; Bernvi, D.C.; Cliff, G.; Wilson, R.P.; Dicken, M.L.; Menzel, J.; Wroe, S.; Pirlo, J.; Pimiento, C. The extinct shark *Otodus megalodon* was a transoceanic superpredator: Inferences from 3D modeling. *Sci. Adv.* **2022**, *8*, eabm9424. [CrossRef]
21. Shimada, K.; Yamaoka, Y.; Kurihara, Y.; Takakuwa, Y.; Maisch IV, H.M.; Becker, M.A.; Eagle, R.A.; Griffiths, M.L. Tessellated calcified cartilage and placoid scales of the Neogene megatooth shark, *Otodus megalodon* (Lamniformes: Otodontidae), offer new insights into its biology and the evolution of regional endothermy and gigantism in the otodontid clade. *Hist. Biol.* **2023**, *in press*. [CrossRef]

22. Pollerspöck, J.; Straube, N. World Wide Web Electronic Publication, Version 2023. 2023. Available online: www.shark-references.com (accessed on 1 September 2023).
23. Yun, C.-G. First Deep-Sea Shark Fossil Teeth from the Miocene of South Korea. *Zoodiversity* **2021**, *55*, 225–232. [CrossRef]
24. Compagno, L.J.V. FAO Species Catalogue. Vol 4: Sharks of the world, Part 1—Hexanchiformes to Lamniformes. *FAO Fish. Synop.* **1984**, *4*, 1–250.
25. Pfeil, F.H. Haie und Rochen aus Walbertsweiler. In *Notizen zu Einem Profil der Selachier-Fundstelle Walbertsweiler im Bereich der miozänen Oberen Meeresmolasse Süddeutschlands*; Barthelt, D., Fejfar, O., Pfeil, F.H., Unger, E., Eds.; Münchner Geowissenschaftliche Abhandlungen, Reihe A; 1991; Volume 19, pp. 198–208. Available online: https://shark-references.com/literature/127 (accessed on 1 September 2023).
26. Hovestadt, D.C. Taxonomic adjustments of the Oligocene and Miocene Odontaspididae and Carchariidae based on extant specimens. *Cainozoic Res.* **2020**, *20*, 229–255.
27. De Schutter, P.J. *Carcharias vorax* (Le Hon, 1871) (Chondrichthyes, Lamniformes), from the Miocene of Belgium: Redescription and designation of a neotype and paraneotype. *Geol. Belg.* **2011**, *14*, 175–192.
28. Reinecke, T.; Louwye, S.; Havekost, U.; Moths, H. The elasmobranch fauna of the late Burdigalian, Miocene, at Werder-Uesen, Lower Saxony, Germany, and its relationships with Early Miocene faunas in the North Atlantic, Central Paratethys and Mediterranean. *Palaeontos* **2011**, *20*, 1–170.
29. Reinecke, T.; Stapf, H.; Raisch, M. Die Selachier und Chimären des Unteren Meeressandes und Schleichsandes im Mainzer Becken (Rupelium, Unteres Oligozän). *Palaeontos* **2001**, *1*, 1–73.
30. Ebert, D.A.; Dando, M.; Fowler, S. *Sharks of the World—A Fully Illustrated Guide*; Wild Nature Press: Princeton, NJ, USA, 2021.
31. Reinecke, T.; von der Hocht, F.; Gille, D.; Kindlimann, R. A review of the odontaspidid shark *Carcharoides* AMEGHINO 1901 (Lamniformes, Odontaspididae) in the Chattian and Rupelian of the North Sea Basin, with the definition of a neotype of *Carcharoides catticus* (PHILIPPI, 1846) and description of a new species. *Palaeontos* **2018**, *31*, 1–75.
32. Purdy, R.W.; Schneider, V.P.; Applegate, S.P.; McLellan, J.H.; Meyer, R.L.; Slaughter, R. The Neogene sharks, rays, and bony fishes from Lee Creek Mine, Aurora, North Carolina. In *Geology and Paleontology of the Lee Creek Mine, North Carolina, III*; Smithsonian Contributions to Paleobiology; Ray, C.E., Bohaska, D.J., Eds.; Smithsonian Institution Press: Washington, DC, USA, 2001; Volume 90, pp. 71–202.
33. Kent, B.W. *Fossil Sharks of the Chesapeake Bay Region*; Egan Rees and Boyer, Inc.: Columbia, SC, USA, 1994.
34. Ebersole, J.A.; Ebersole, S.M.; Cicimurri, D.J. The occurrence of early Pleistocene marine fish remains from the Gulf Coast of Mobile County, Alabama, USA. *Palaeodiversity* **2017**, *10*, 97–115. [CrossRef]
35. Whitenack, L.B.; Gottfried, M.D. A Morphometric Approach for Addressing Tooth-Based Species Delimitation in Fossil Mako Sharks, *Isurus* (Elasmobranchii: Lamniformes). *J. Vertebr. Paleontol.* **2010**, *30*, 17–25. [CrossRef]
36. Cione, A.L.; Cabrera, D.A.; Barla, M.J. Oldest record of the Great White Shark (Lamnidae, *Carcharodon*; Miocene) in the Southern Atlantic. *Geobios* **2012**, *45*, 167–172. [CrossRef]
37. Collareta, A.; Landini, W.; Chacaltana, C.; Valdivia, W.; Altamirano-Sierra, A.; Urbina-Schmitt, M.; Bianucci, G. A well preserved skeleton of the fossil shark *Cosmopolitodus hastalis* from the late Miocene of Peru, featuring fish remains as fossilized stomach contents. *Riv. Ital. Paleontol. Stratigr.* **2017**, *123*, 11–22. [CrossRef]
38. Takakuwa, Y. A dense occurrence of teeth of fossil "mako" shark ("Isurus" hastalis: Chondrichthyes, Lamniformes), associated with a balaenopterid-whale skeleton of the Late Miocene Pisco Formation, Peru, South America. *Bull. Gunma Mus. Nat. Hist.* **2014**, *18*, 77–86.
39. Bianucci, G.; Sorce, B.; Storai, T.; Landini, W. Killing in the Pliocene: Shark attack on a dolphin from Italy. *Palaeontology* **2010**, *53*, 457–470. [CrossRef]
40. Höltke, O.; Rasser, M.W.; Unger, E.; Pollerspöck, J. The Elasmobranch Fauna from the Upper Marine Molasse (Lower Miocene, Burdigalian) of Ursendorf (District Sigmaringen, Baden-Württemberg, SW-Germany). *Palaeontos* **2020**, *33*, 1–56.
41. Boessenecker, R.W. First record of the megatoothed shark *Carcharocles megalodon* from the Mio-Pliocene Purisima Formation of Northern California. *PaleoBios* **2016**, *33*, 1–7. [CrossRef]
42. Hoffmeister, C.; Felipe, M.; Villafaña, J.A. The Neogene Record of Cartilaginous Fishes (Chondrichthyes: Holocephali, Elasmobranchii) from Northern Chile: A Review and Identification Guide. *J. S. Am. Earth Sci.* **2023**, *124*, 104230. [CrossRef]
43. Shimada, K.; Chandler, R.E.; Lam, O.L.T.; Tanaka, T.; Ward, D.J. A new elusive otodontid shark (Lamniformes: Otodontidae) from the lower Miocene, and comments on the taxonomy of otodontid genera, including the 'megatoothed' clade. *Hist. Biol.* **2016**, *29*, 704–714. [CrossRef]
44. Landini, W.; Collareta, A.; Di Celma, C.; Malinverno, E.; Urbina, M.; Bianucci, G. The early Miocene elasmobranch assemblage from Zamaca (Chilcatay Formation, Peru). *J. S. Am. Earth Sci.* **2019**, *91*, 352–371. [CrossRef]
45. Perez, V.J.; Godfrey, S.J.; Kent, B.W.; Weems, R.E.; Nance, J.R. The transition between *Carcharocles chubutensis* and *Carcharocles megalodon* (Otodontidae, Chondrichthyes): Lateral cusplet loss through time. *J. Vertebr. Paleontol.* **2019**, *38*, e1546732. [CrossRef]
46. Pollerspöck, J.; Güthner, T.; Straube, N. Re-evaluation of the Lower Miocene (Burdigalian, Ottnangian) elasmobranch fauna (Elasmobranchii, Neoselachii) from Upper Austria (Allerding, near Schärding, Austria) with comments on the paleogeographic distribution of the recorded squaliform sharks. *Ann. Naturhistorischen Mus. Wien Ser. A* **2022**, *122*, 87–163.

47. Shimada, K.; Maisch, H.M.; Perez, V.J.; Becker, M.A.; Griffiths, M.L. Revisiting body size trends and nursery areas of the Neogene megatooth shark, *Otodus megalodon* (Lamniformes: Otodontidae), reveals Bergmann's rule possibly enhanced its gigantism in cooler waters. *Hist. Biol.* **2022**, *35*, 208–217. [CrossRef]
48. Pimiento, C.; Ehret, D.J.; Macfadden, B.J.; Hubbell, G. Ancient Nursery Area for the Extinct Giant Shark Megalodon from the Miocene of Panama. *PLoS ONE* **2010**, *5*, e10552. [CrossRef] [PubMed]
49. Herraiz, J.L.; Ribé, J.; Botella, H.; Martínez-Pérez, C.; Ferrón, H.G. Use of nursery areas by the extinct megatooth shark *Otodus megalodon* (Chondrichthyes, Lamniformes). *Biol. Lett.* **2020**, *16*, 20200746. [CrossRef]
50. Blackburn, D.G. Evolution of vertebrate viviparity and specializations for fetal nutrition: A quantitative and qualitative analysis. *J. Morphol.* **2015**, *276*, 961–990. [CrossRef]
51. Shimada, K.; Bonnan, M.F.; Becker, M.A.; Griffiths, M.L. Ontogenetic growth pattern of the extinct megatooth shark *Otodus megalodon*-implications for its reproductive biology, development, and life expectancy. *Hist. Biol.* **2021**, *33*, 3254–3259. [CrossRef]
52. Gottfried, M.D.; Compagno, L.J.V.; Bowman, S.C. Size and Skeletal anatomy of the Giant Megatooth Shark *Carcharodon megalodon*. In *Great White Sharks. The Biology of Carcharodon carcharias*; Klimley, A.P., Ainley, Eds.; Academic Press: Cambridge, MA, USA, 1996; pp. 55–66. [CrossRef]
53. Collareta, A.; Lambert, O.; Landini, W.; Di Celma, C.; Malinverno, E.; Varas-Malca, R.; Urbina, M.; Bianucci, G. Did the giant extinct shark *Carcharocles megalodon* target small prey? Bite marks on marine mammal remains from the late Miocene of Peru. *Palaeogeogr. Palaeoclimatol. Palaeoecol.* **2017**, *469*, 84–91. [CrossRef]
54. Purdy, R.W. Paleoecology of fossil White Sharks. In *Great White Sharks. The Biology of Carcharodon carcharias*; Klimley, A.P., Ainley, D., Eds.; Academic Press: San Diego, CA, USA, 1996; pp. 133–139.
55. Griffiths, M.L.; Eagle, R.A.; Kim, S.L.; Flores, R.J.; Becker, M.A.; Maisch, H.M.; Trayler, R.B.; Chan, R.L.; McCormack, J.; Akhtar, A.A.; et al. Endothermic physiology of extinct megatooth sharks. *Proc. Natl. Acad. Sci. USA* **2023**, *120*, e2218153120. [CrossRef]
56. Pimiento, C.; Clements, C.F. When Did Carcharocles megalodon Become Extinct? A New Analysis of the Fossil Record. *PLoS ONE* **2014**, *9*, e111086. [CrossRef]
57. Boessenecker, R.W.; Ehret, D.J.; Long, D.J.; Churchill, M.; Martin, E.; Boessenecker, S.J. The Early Pliocene extinction of the mega-toothed shark *Otodus megalodon*: A view from the eastern North Pacific. *PeerJ* **2019**, *7*, e6088. [CrossRef] [PubMed]
58. Pimiento, C.; Macfadden, B.J.; Clements, C.F.; Varela, S.; Jaramillo, C.; Velez-Juarbe, J.; Silliman, B.R. Geographical distribution patterns of *Carcharocles megalodon* over time reveal clues about extinction mechanisms. *J. Biogeogr.* **2016**, *43*, 1645–1655. [CrossRef]
59. Kent, B.W.; Powell, G.W. Reconstructed dentition of the rare lamnoid shark *Parotodus benedeni* (Le Hon) from the Yorktown Formation (Early Pliocene) at Lee Creek Mine, North Carolina. *Mosasaur* **1999**, *6*, 1–10.
60. Collareta, A.; Casati, S.; Di Cencio, A. The Palaeobiology of the False Mako Shark, *Parotodus benedenii* (Le Hon, 1871): A View from the Pliocene Mediterranean Sea. *J. Mar. Sci. Eng.* **2023**, *11*, 1990. [CrossRef]
61. Welton, B.J. A New Archaic Basking Shark (Lamniformes: Cetorhinidae) from the Late Eocene of Western Oregon, U.S.A., and Description of the Dentition, Gill Rakers and Vertebrae of the Recent Basking Shark *Cetorhinus maximus* (Gunnerus). *New Mex. Mus. Nat. Hist. Sci. Bull.* **2013**, *58*, 1–48.
62. Hovestadt, D.C.; Hovestadt-Euler, M. A partial skeleton of *Cetorhinus parvus* Leriche, 1910 (Chondrichthyes, Cetorhinidae) from the Oligocene of Germany. *Paläontologische Z.* **2011**, *86*, 71–83. [CrossRef]
63. Krak, A.M.; Shimada, K. The dentition of the extinct megamouth shark, *Megachasma applegatei* (Lamniformes: Megachasmidae), from southern California, USA, based on geometric morphometrics. *PaleoBios* **2023**, *40*, P940160139. [CrossRef]
64. Shimada, K.; Welton, B.J.; Long, D.J. A new fossil megamouth shark (Lamniformes, Megachasmidae) from the Oligocene-Miocene of the western United States. *J. Vertebr. Paleontol.* **2014**, *34*, 281–290. [CrossRef]
65. Shimada, K. Dental homologies in lamniform sharks (Chondrichthyes: Elasmobranchii). *J. Morphol.* **2002**, *251*, 38–72. [CrossRef]
66. Collareta, A.; Merella, M.; Mollen, F.H.; Casati, S.; Di Cencio, A. The extinct catshark *Pachyscyllium distans* (Probst, 1879) (Elasmobranchii: Carcharhiniformes) in the Pliocene of the Mediterranean Sea. *Neues Jahrbuch für Geologie und Paläontologie, Abhandlungen* **2020**, *295*, 129–139. [CrossRef]
67. Ward, D.J.; Bonavia, C.G. Additions to, and a review of, the Miocene shark and ray fauna of Malta. *Cent. Mediterr. Nat.* **2001**, *3*, 131–146.
68. Pimiento, C.; Cantalapiedra, J.L.; Shimada, K.; Field, D.J.; Smaers, J.B. Evolutionary pathways toward gigantism in sharks and rays. *Evolution* **2019**, *73*, 588–599. [CrossRef] [PubMed]
69. Feichtinger, I.; Pollerspöck, J. *Haie im Alpenvorland—Fossile Zeugen eines Verschwundenen Paradieses*; Verlag Anton Pustet: Salzburg, Austria, 2021.
70. Türtscher, J.; López-Romero, F.A.; Jambura, P.L.; Kindlimann, R.; Ward, D.J.; Kriwet, J. Evolution, diversity, and disparity of the tiger shark lineage *Galeocerdo* in deep time. *Paleobiology* **2021**, *47*, 574–590. [CrossRef] [PubMed]
71. Feichtinger, I.; Fritz, I.; Göhlich, U.B. Tiger shark feeding on sirenia—First fossil evidence from the middle Miocene of the Styrian Basin (Austria). *Hist. Biol.* **2021**, *34*, 193–200. [CrossRef]
72. Godfrey, S.J.; Smith, J.B. Shark-bitten vertebrate coprolites from the Miocene of Maryland. *Naturwissenschaften* **2010**, *97*, 461–467. [CrossRef] [PubMed]
73. Collareta, A.; Kindlimann, R.; Baglioni, A.; Landini, W.; Sarti, G.; Altamirano, A.; Urbina, M.; Bianucci, G. Dental Morphology, Palaeoecology and Palaeobiogeographic Significance of a New Species of Requiem Shark (Genus Carcharhinus) from the Lower Miocene of Peru (East Pisco Basin, Chilcatay Formation). *J. Mar. Sci. Eng.* **2022**, *10*, 1466. [CrossRef]

74. Collareta, A.; Merella, M.; Casati, S.; Di Cencio, A.; Bianucci, G. Smoking guns for cold cases: The find of a Carcharhinus tooth piercing a fossil cetacean rib, with notes on the feeding ecology of some Mediterranean Pliocene requiem sharks. *Neues Jahrb. Geol. Paläontologie Abh.* **2022**, *305*, 145–152. [CrossRef]
75. Merella, M.; Collareta, A.; Casati, S.; Di Cencio, A.; Bianucci, G. An unexpected deadly meeting: Deep-water (hexanchid) shark bite marks on a sirenian skeleton from Pliocene shoreface deposits of Tuscany (Italy). *Neues Jahrb. Geol. Paläontologie Abh.* **2021**, *301*, 295–305, Erratum in *Neues Jahrb. Geol. Paläontologie Abh.* **2022**, *303*, 239–241. [CrossRef] [PubMed]
76. Laurito, C.A.; Calvo, C.; Valerio, A.L.; Calvo, A.; Chacón, R. Ictiofauna del mioceno inferior de la localidad de Pacuare de Tres Equis, formación río Banano, provincia de Cartago, Costa Rica, y descripción de un nuevo género y una nueva especie de scaridae. *Rev. Geológica América Cent.* **2014**, *50*, 153–192. [CrossRef]
77. Cigala-Fulgosi, F. Heptranchias perlo (Bonnaterre) (Selachii, Hexanchidae) nel Serravalliano di Visiano (Medesano, Parma, Emilia Occidentale): Considerazioni tassonomiche e filogenetiche. *Boll. Soc. Paleontol. Ital.* **1977**, *16*, 245–256.
78. Carnevale, G. Fossil fishes from the Serravallian (Middle Miocene) of Torricella Peligna, Italy. *Palaeontogr. Ital.* **2005**, *91*, 1–67.
79. Carrillo-Briceño, J.D.; de Gracia, C.; Pimiento, C.; Aguilera, O.A.; Kindlimann, R.; Santamarina, P.; Jaramillo, C. A New Late Miocene Chondrichthyan Assemblage from the Chagres Formation, Panama. *J. S. Am. Earth Sci.* **2015**, *60*, 56–70. [CrossRef]
80. Antunes, M.T.; Jonet, S. Requins de l'Helvétien supérieur et du Tortonien de Lisbonne. *Rev. Fac. Ciências Univ. Lisb.* **1970**, *16*, 119–280.
81. Aguilera, O.A.; de Aguilera, D.R. An exceptional coastal upwelling fish assemblage in the Caribbean neogene. *J. Paleontol.* **2001**, *75*, 732–742. [CrossRef]
82. Ledoux, J.-C. Les Squalidae (Euselachii) miocènes des environs d'Avignon (Vaucluse). *Doc. Lab. Geol. Fac. Sci. Lyon Notes Mém.* **1972**, *52*, 133–175.
83. Landini, W. Revizione degli «Ittiodontoliti pliocenici» della collezione Lawley. *Palaeontogr. Ital.* **1977**, *70*, 92–134.
84. Cappetta, H.; Nolf, D. Les sélaciens du Pliocène inférieur de Le-Puget-sur-Argens (Sud-Est de la France). *Palaeontographica* **1991**, *218*, 49–67.
85. Cigala-Fulgosi, F.; Casati, S.; Orlandini, A.; Persico, D. A small fossil fish fauna, rich in *Chlamydoselachus* teeth, from the Late Pliocene of Tuscany (Siena, central Italy). *Cainozoic Res.* **2009**, *6*, 3–23.
86. Carrillo-Briceño, J.D.; Villafaña, J.A.; de Gracia, C.; Flores-Alcívar, F.F.; Kindlimann, R.; Abella, J. Diversity and paleoenvironmental implications of an elasmobranch assemblage from the Oligocene–Miocene boundary of Ecuador. *PeerJ* **2020**, *8*, e9051. [CrossRef] [PubMed]
87. Höltke, O.; Maxwell, E.E.; Bracher, H.; Rasser, M.W. The shark and ray teeth of the Lower Miocene (Upper Marine Molasse) from Ballendorf, Baden-Württemberg, Southern Germany. *Palaeobiodiversity Palaeoenvironments* **2023**, *in press*. [CrossRef]
88. Martínez-Pérez, C.; Carrillo-Briceño, J.D.; Esparza, C.; Ferrón, H.G.; Manzanares, E.; Hammann, C.; Botella, H. A Serravallian (Middle Miocene) shark fauna from Southeastern Spain and its palaeoenvironment significance. *Hist. Biol.* **2018**, *30*, 422–432. [CrossRef]
89. Suzuki, H. A fossil deep-sea shark assemblage from the Middle Miocene, Ueda City, Nagano Prefecture, central Japan. *Earth Sci. Chikyu Kagaku* **2012**, *66*, 47–61. (In Japanese)
90. Cigala-Fulgosi, F. A deep water elasmobranch fauna from a lower Pliocene outcropping (Northern Italy). In *Proceedings of the Second International Conference on Indo-Pacific Fishes*; Uyeno, T., Arai, R., Taniuchi, T., Matsuura, K., Eds.; Ichthyological Society of Japan: Tokyo, Japan, 1986; pp. 133–139.
91. Comaschi, C.I. *I Pesci del Miocene della Sardigna*; Stabilimento Tipografico Editoriale Fossataro: Cagliari, Italy, 1973.
92. Cappetta, H. Les Sélaciens miocènes du Midi de la France. Répartitions stratigraphique et bathymétrique. *3ème Réunion Annu. Sci. Terre* **1975**, *90*.
93. Brisswalter, G. Inventaire des Elasmobranches (requins, raies, chimères) des dépôts molassiques du Sud-Luberon (Miocène supérieur), à Cabrières d'Aigues (Vaucluse) France. *Courr. Sci. Parc Régional Lubéron* **2009**, 1–100.
94. Carrillo-Briceño, J.D.; Argyriou, T.; Zapata, V.; Kindlimann, R.; Jaramillo, C. A new early Miocene (Aquitanian) Elasmobranchii assemblage from the La Guajira Peninsula, Colombia. *Ameghiniana* **2016**, *53*, 77–99. [CrossRef]
95. Carrillo-Briceño, J.D.; Luz, Z.; Hendy, A.; Kocsis, L.; Aguilera, O.; Vennemann, T. Neogene Caribbean elasmobranchs: Diversity, paleoecology and paleoenvironmental significance of the Cocinetas Basin assemblage (Guajira Peninsula, Colombia). *Biogeosciences* **2019**, *16*, 33–56. [CrossRef]
96. Itoigawa, J.; Nishimoto, N.; Karasawa, H.; Okumura, Y. Miocene fossils of the Mizunami group, central Japan. 3. Elasmobranchs. *Monogr. Mizunami Foss. Mus.* **1985**, *5*, 1–99.
97. Yabe, H.; Hirayama, R. Selachian fauna from the Upper Miocene Senhata Formation, Boso Peninsula, Central Japan. *Nat. Hist. Res.* **1998**, *5*, 33–61.
98. Tanaka, T. Fossil elasmobranchs from the Miura Group (Late Miocene-Early Pliocene) in the Miura Peninsula, Central Japan. [in Japanese, with English summary]. *Bull. Kanagawa Prefect. Mus. Nat. Sci.* **2001**, *22*, 73–80.
99. Suzuki, H. On the fossil teeth of kitefin shark from the Middle Miocene Iseyama Formation, Sanada-machi, Nagano Prefecture, central Japan. *Earth Sci. Chikyu Kagaku* **2005**, *59*, 383–388. (In Japanese)
100. Uyeno, T.; Matsushima, Y. Pliocene shark remains of *Carcharodon, Carcharhinus, and Dalatias*, from Kanagawa Prefecture, Japan. *Bull. Kanagawa Prefect. Mus. Nat. Sci.* **1975**, *8*, 41–55.

101. Cigala-Fulgosi, F. Rare oceanic deep water squaloid sharks from the Lower Pliocene of the Northern Apennines (Parma province, Italy). *Boll. Soc. Paleontol. Ital.* **1996**, *34*, 301–322.
102. Marsili, S.; Tabanelli, C. Bathyal sharks from the middle Pliocene of the Romagna Apennines (Italy). *Neues Jahrb. Geol. Paläontologie Abh.* **2007**, *244*, 247–255. [CrossRef]
103. Cigala-Fulgosi, F. Addition to the Pliocene fish fauna of Italy. Evidence of *Somniosus rostratus* (Risso, 1826) from the foothills of the Northern Apennines (Parma Province, Italy) (Chondrichthyes, Squalidae). *Tert. Res.* **1988**, *10*, 101–106.
104. Cicimurri, D.J.; Knight, J.L. Late Oligocene sharks and rays from the Chandler Bridge Formation, Dorchester County, South Carolina, USA. *Acta Palaeontol. Pol.* **2009**, *54*, 627–647. [CrossRef]
105. Cappetta, H. Les Sélaciens du Miocène de la région de Montpellier. *Palaeovertebrata Mémoire Extraordin.* **1970**, 1–139. [CrossRef]
106. Visaggi, C.C.; Godfrey, S.J. Variation in Composition and Abundance of Miocene Shark Teeth from Calvert Cliffs, Maryland. *J. Vertebr. Paleontol.* **2010**, *30*, 26–35. [CrossRef]
107. Laurito, C.A. *Los Selaceos Fosiles de la Localidad de Alto Guayacan (y Otros Ictiolitos Associados)*; Mioceno Superior-Plioceno Inferior de Limon; Guila Imprenta: San José, Costa Rica, 1999.
108. Welton, B.J. *Cetorhinus* cf. *C. maximus* (Gunnerus) (Lamniformes: Cetorhinidae), A Basking Shark from the Late Miocene Empire Formation, Coos Bay, Oregon. *Bull. South. Calif. Acad. Sci.* **2013**, *112*, 74–92. [CrossRef]
109. Uyeno, T.; Ono, K.; Sakamoto, O. Miocene Elasmobranchs from Chichibu Basin, Saitama, Japan. *Bull. Saitama Mus. Nat. Hist.* **1983**, *1*, 27–36.
110. Schultz, O.; Brzobohatý, R.; Kroupa, O. Fish teeth from the Middle Miocene of Kienberg at Mikulov, Czech Republic, Vienna Basin. *Ann. Naturhistorischen Mus. Wien Ser. A* **2010**, *112*, 489–506.
111. Powell, C.L.; Boessenecker, R.W.; Smith, N.A.; Fleck, R.J.; Carlson, S.J.; Allen, J.R.; Long, D.J.; Sarna-Wojcicki, A.M.; Guruswami-Naidu, R.B. *Geology and Paleontology of the Late Miocene Wilson Grove Formation at Bloomfield Quarry, Sonoma County, California*; U.S. Geological Survey Scientific Investigations Report 2019–5021; U.S. Geological Survey: Reston, VA, USA, 2019; p. 77. [CrossRef]
112. Lienau, H.-W. Haie und Rochen aus dem Sylter Obermiozän. In *Fossilien von Sylt II*; von Hacht, U., Ed.; Verlag Inge–Maria von Hacht: Hamburg, Germany, 1987; pp. 19–95.
113. Menzel, H.; George, P.; Schieck, H. Eine miozäne Scholle aus der Tongrube in Rhaden bei Lamstedt. *Aufschluss* **1994**, *45*, 11–25.
114. Long, D.J. Late Miocene and Early Pliocene fish assemblages from the north coast of Chile. *Tert. Res.* **1993**, *14*, 117–126.
115. Van den Bosch, M.; Cadée, M.C.; Janssen, A.W. Lithostratigraphical and biostratigraphical subdivision of Tertiary deposits (Oligocene—Pliocene) in the Winterswijk—Almelo region (eastern part of the Netherlands). *Scr. Geol.* **1975**, *29*, 1–167.
116. Herman, J. Réflexions sur la systématique des Galeoidei et sur les affinités du genre *Cetorhinus* à l'occasion de la découverte d'éléments de la denture d'un exemplaire fossile dans les sables du Kattendijk à Kallo (Pliocène inférieur, Belgique). *Ann. Société Géologique Belg.* **1979**, *102*, 357–377.
117. Leriche, M. Sur un appareil fanonculaire de *Cetorhinus* trouvé à l'état fossile dans le Pliocène d'Anvers. *Comptes Rendus Hebd. Séances L'academie Sci.* **1908**, *146*, 875–878.
118. Cione, A.L.; Mennucci, J.; Santalucita, F.; Acosta Hospitaleche, C. Local extinction of sharks of genus *Carcharias* Rafinesque, 1810 (Elasmobranchii, Odontaspididae) in the eastern Pacific Ocean. *Rev. Geológica Chile* **2007**, *34*, 139–145. [CrossRef]
119. Suárez, M.E.; Encinas, A.; Ward, D.J. An Early Miocene elasmobranch fauna from the Navidad Formation, Central Chile, South America. *Cainozoic Res.* **2006**, *4*, 3–18.
120. Schultz, O. Pisces. In *Catalogus Fossilium Austriae, Band 3*; Piller, W.E., Ed.; Verlag der Österreichischen Akademie der Wissenschaften: Wien, Austria, 2013.
121. Kordos, L.; Solt, P. A magyarországi Miocén tengeri gerinces faunaszintek vázlata. [in Hungarian: Sketch of the marine vertebrate fauna levels of the Miocene of Hungary]. In *A Magyar Állami Földtani Intézet évi Jelentése*; Magyar Állami Földtani Intézet: Budapest, Hungary, 1984; pp. 347–351.
122. Schalch, F. Ueber einige Tertiärbildungen der Umgebung von Schaffhausen. *Neues Jahrbuch Mineralogie, Geologie Paläontologie* **1881**, *2*, 42–76.
123. Cigala-Fulgosi, F. Addition to the fish fauna of the Italian Miocene. The occurrence of *Pseudocarcharias* (Chondrichthyes, Pseudocarchariidae) in the lower Serravallian of Parma Province, Northern Apennines. *Tert. Res.* **1992**, *14*, 51–60.
124. Antunes, M.T.; Balbino, A.C.; Cappetta, H. Sélaciens du Miocène terminal du Bassin d'Alvalade (Portugal). Essai de synthèse. *Ciências Terra (UNL)* **1999**, *13*, 115–129.
125. Pfeil, F.H. Eine nektonische Fischfauna aus dem unteroligozänen Schönecker Fischschiefer des Galon-Grabens in Oberbayern. *Geol. Bavarica* **1981**, *82*, 357–388.
126. Case, G.R. A selachian fauna from the Trent Formation, Lower Miocene (Aquitanian) of Eastern North Carolina. *Palaeontographica* **1980**, *171*, 75–103.
127. Bor, T.J.; Reinecke, T.; Verschueren, S. Miocene Chondrichthyes from Winterswijk—Miste, the Netherlands. *Palaeontos* **2012**, *21*, 1–136.
128. Boyd, B.M. *Fossil Sharks and Rays of Gainesville Creeks; Alachua COUNTY, Florida: Hogtown Group; (Middle Miocene to Lower Pliocene)*; Florida Paleontological Society: Gainesville, FL, USA, 2016.
129. Perez, V.J.; Pimiento, C.; Hendy, A.; González-Barba, G.; Hubbell, G.; Macfadden, B.J. Late Miocene chondrichthyans from Lago Bayano, Panama: Functional diversity, environment and biogeography. *J. Paleontol.* **2017**, *91*, 512–547. [CrossRef]

130. Balbino, A.C. Sharks from the Middle and early Upper Miocene from Lisbon, Portugal. A check-list. *Comun. Inst. Geológico Min.* **1996**, *82*, 141–144.
131. Antunes, M.T.; Balbino, A.C. Uppermost Miocene Lamniform Selachians (Pisces) from the Alvalade Basin (Portugal). *Ciências Terra (UNL)* **2003**, *15*, 141–154.
132. Cigala-Fulgosi, F. Additions to the Eocene and Pliocene fish fauna of Italy. Evidence of *Alopias* cf. *denticulatus* Cappetta, 1981 in the Bartonian-Priabonian of the Monte Piano Marl (Northern Apennines) and of *A. superciliosus* (Lowe, 1840) in the Pliocene of Tuscany (Chondrichthyes, Alopiidae). *Tert. Res.* **1988**, *10*, 93–99.
133. Noetling, F. Fauna of the Miocene beds of Burma. *Mem. Geol. Surv. India Palaeontol. Indica New Ser.* **1901**, *1*, 1–378.
134. Bhalla, S.N.; Dev, P. A Preliminary Note on Miocene Elasmobranchs from Orissa. *J. Geol. Soc. India* **1975**, *16*, 98–99.
135. Antunes, M.T.; Jonet, S.; Nascimento, A. Vertébrés (crocodiliens, poissons) du Miocène marin de l'Argarve occidentale. *Ciências Terra (UNL)* **1981**, *6*, 9–38.
136. Reinecke, T.; Moths, H.; Grant, A.; Breitkreuz, H. Die Elasmobranchier des norddeutschen Chattiums, insbesondere des Sternberger Gesteins (Eochattium, Oligozän). *Palaeontos* **2005**, *8*, 1–135.
137. Mollen, F.H. A partial rostrum of the porbeagle shark *Lamna nasus* (Lamniformes, Lamnidae) from the Miocene of the North Sea Basin and the taxonomic importance of rostral morphology in extinct sharks. *Geol. Belg.* **2010**, *13*, 61–76.
138. Collareta, A.; Casati, S.; Di Cencio, A. The porbeagle shark, *Lamna nasus* (Elasmobranchii: Lamniformes), from the late Pliocene of the central Mediterranean Basin. *Neues Jahrb. Geol. Paläontologie Abh.* **2018**, *287*, 307–316. [CrossRef]
139. Itoigawa, J. Miocene palaeogeography of the Mizunami Group of the Tono region, central Japan. *Palaeogeogr. Palaeoclimatol. Palaeoecol.* **1993**, *100*, 209–215. [CrossRef]
140. Domeier, M.L. *Global Perspectives on the Biology and Life History of the White Shark*; CRC Press: Boca Raton, FL, USA, 2012.
141. Cigala-Fulgosi, F. Predation (or possible scavenging) by a great white shark on an extinct species of bottlenose dolphin in the Italian Pliocene. *Tert. Res.* **1990**, *12*, 17–36.
142. Bianucci, G.; Bisconti, M.; Landini, W.; Storai, T.; Zuffa, M.; Giuliani, S.; Mojetta, A. Trophic interaction between white shark, *Carcharodon carcharias*, and cetaceans: A comparison between Pliocene and recent data from central Mediterranean Sea. In Proceedings of the 4th European Elasmobranch Association Meeting, Livorno, Italy, 27–30 September 2000; pp. 33–48.
143. Adnet, S.; Balbino, A.C.; Antunes, M.T.; Marín-Ferrer, J.M. New fossil teeth of the White Shark (*Carcharodon carcharias*) from the Early Pliocene of Spain. Implication for its paleoecology in the Mediterranean. *Neues Jahrb. Für Geol. Und Paläontologie Abh.* **2010**, *256*, 7–16. [CrossRef]
144. Villafaña, J.A.; Hernandez, S.; Alvarado, A.; Shimada, K.; Pimiento, C.; Rivadeneira, M.M.; Kriwet, J. First evidence of a palaeo-nursery area of the great white shark. *Sci. Rep.* **2020**, *10*, 8502. [CrossRef]
145. Case, G.R.; Borodin, P.D.A. Middle Eocene Selachian Fauna from the Castle Hayne Limestone Formation of Duplin County, North Carolina. *Münchner Geowiss. Abh. Reihe A Geol. Paläontologie* **2000**, *39*, 17–32.
146. Pledge, N.S. *An Early Pliocene Shark Tooth Assemblage in South Australia*; South Australia Department of Mines and Energy, Special Publication: Adelaide, Australia, 1985; Volume 5, pp. 287–299.
147. Fitch, J.E.; Reimer, R.D. Otoliths and other fish remains from a Long Beach, California, Pliocene deposit. *Bull. South. Calif. Acad. Sci.* **1967**, *66*, 77–91.
148. Carrillo-Briceño, J.D.; González-Barba, G.; Landaeta, M.F.; Nielsen, S.N. Condrictios fósiles del Plioceno Superior de la Formación Horcón, Región de Valparaíso, Chile central. [Fossil Chondrichthyans from the Upper Pliocene Horcón Formation, Valparaíso Region, central Chile]. *Rev. Chil. Hist. Nat.* **2013**, *86*, 191–206. [CrossRef]
149. Sharma, K.M.; Singh, N.A.; Patnaik, R.; Tiwari, R.P.; Singh, N.P.; Singh, Y.P.; Choudhary, D.; Lalotra, S.K. Sharks and rays (chondrichthyes, elasmobranchii) from the miocene sediments of Kutch, Gujarat, India: Paleoenvironmental and paleobiogeographic implications. *Hist. Biol.* **2021**, *34*, 10–29. [CrossRef]
150. Szabó, M.; Kocsis, L.; Szabó, P.; Békési, Z.; Gulyás, P. New records and specimens to the Badenian fish fauna of Nyirád (Hungary), including the first report of *Galeocerdo cuvier* from the Middle Miocene of Europe. *Fragm. Palaeontol. Hung.* **2023**, *38*, 53–74. [CrossRef]
151. Carrillo-Briceño, J.D.; Maxwell, E.; Aguilera, O.A.; Sánchez, R.; Sánchez-Villagra, M.R. Sawfishes and Other Elasmobranch Assemblages from the Mio-Pliocene of the South Caribbean (Urumaco Sequence, Northwestern Venezuela). *PLoS ONE* **2015**, *10*, e0139230. [CrossRef]
152. Alberti, M.; Reich, S. A palaeoecological review of the lower Gatun Formation (Miocene) of Panama with special emphasis on trophic relationships. *Palaeobiodiversity Palaeoenvironments* **2018**, *98*, 571–591. [CrossRef]
153. Kocsis, L.; Botfalvai, G.; Qamarina, Q.; Razak, H.; Kiraly, E.; Lugli, F.; Wings, O.; Lambertz, M.; Raven, H.; Briguglio, A.; et al. Geochemical analyses suggest stratigraphic origin and late Miocene age of reworked vertebrate remains from Penanjong Beach in Brunei Darussalam (Borneo). *Hist. Biol.* **2021**, *33*, 2627–2638. [CrossRef]
154. Webb, S.D.; Tessmann, N. A Pliocene vertebrate fauna from low elevation in Manatee County, Florida. *Am. J. Sci.* **1968**, *266*, 777–811. [CrossRef]
155. Maisch, H.M.; Becker, M.A.; Chamberlain, J.A. Lamniform and Carcharhiniform Sharks from the Pungo River and Yorktown Formations (Miocene–Pliocene) of the Submerged Continental Shelf, Onslow Bay, North Carolina, USA. *Copeia* **2018**, *106*, 353–374. [CrossRef]

156. Antunes, M.T. Faunes ichthyologiques du Néogène supérieur d'Angola, leur âge, remarques sur le Pliocène marin en Afrique australe. *Ciências Terra (UNL)* **1978**, *4*, 59–90.
157. Pawellek, T.; Adnet, S.; Cappetta, H.; Metais, E.; Salem, M.; Brunet, M.; Jaeger, J.-J. Discovery of an earliest Pliocene relic tropical fish fauna in a newly detected cliff section (Sabratah Basin, NW Libya). *Neues Jahrb. Geol. Paläontologie Abh.* **2012**, *266*, 93–114. [CrossRef]
158. Collareta, A.; Mollen, F.H.; Merella, M.; Casati, S.; Di Cencio, A. Remarkable multicuspid teeth in a new elusive skate (Chondrichthyes, Rajiformes) from the Mediterranean Pliocene. *Paläontologische Z.* **2021**, *95*, 117–128. [CrossRef]
159. Kocsis, L.; Razak, H.; Briguglio, A.; Szabó, M. First report on a diverse Neogene cartilaginous fish fauna from Borneo (Ambug Hill, Brunei Darussalam). *J. Syst. Palaeontol.* **2019**, *17*, 791–819. [CrossRef]
160. Carrillo-Briceño, J.D.; Aguilera, O.A.; Rodriguez, F. Fossil Chondrichthyes from the central eastern Pacific Ocean and their paleoceanographic significance. *J. S. Am. Earth Sci.* **2014**, *51*, 76–90. [CrossRef]
161. Landini, W.; Collareta, A.; Bianucci, G. The origin of biogeographic segregation in the copper shark (*Carcharhinus brachyurus*): An integrative reconstruction based on neontological and paleontological data. *Vie Milieu Life Environ.* **2020**, *70*, 117–132.
162. Landini, W.; Collareta, A.; Pesci, F.; Di Celma, C.; Urbina, M.; Bianucci, G. A secondary nursery area for the copper shark *Carcharhinus brachyurus* from the late Miocene of Peru. *J. S. Am. Earth Sci.* **2017**, *78*, 164–174. [CrossRef]
163. Singh, N.A.; Choudhary, D.; Singh, Y.P.; Singh, N.P.; Patnaik, R.; Tiwari, R.P.; Sharma, K.M. Chondrichthyan and osteichthyan fauna from the middle Miocene deposits of Palasava, Kutch, India: Implication for paleoenvironment and paleobiogeography. *Comptes Rendus Palevol* **2022**, *21*, 939–968. [CrossRef]
164. Pimiento, C.; González-Barba, G.; Ehret, D.J.; Hendy, A.J.W.; Macfadden, B.J.; Jaramillo, C. Sharks and Rays (Chondrichthyes, Elasmobranchii) from the Late Miocene Gatun Formation of Panama. *J. Paleontol.* **2013**, *87*, 755–774. [CrossRef]
165. Carnevale, G.; Marsili, S.; Caputo, D.; Egisti, L. The Silky Shark *Carcharhinus falciformis* (Bibron, 1841) in the Pliocene of Cava Serredi (Fine Basin, Italy). *Neues Jahrb. Geol. Paläontologie Abh.* **2006**, *242*, 357–370. [CrossRef]
166. da Silva Rodrigues-Filho, L.F.; da Costa Nogueira, P.; Sodré, D.; da Silva Leal, J.R.; Nunes, J.L.S.; Rincon, G.; Lessa, R.P.T.; Sampaio, I.; Vallinoto, M.; Ready, J.S.; et al. Evolutionary History and Taxonomic Reclassification of the Critically Endangered Daggernose Shark, a Species Endemic to the Western Atlantic. *J. Zool. Syst. Evol. Res.* **2023**, *2023*, 4798805. [CrossRef]
167. Deraniyagala, P.E.P. A Miocene vertebrate faunule from the Malu Member of Ceylon. *Spolia Zeylan* **1969**, *31*, 551–570.
168. Goolaerts, S.; de Ceuster, J.; Mollen, F.H.; Gijsen, B.; Bosselaers, M.; Lambert, O.; Uchman, A.; Van Herck, M.; Adriaens, R.; Houthuys, R.; et al. The upper Miocene Deurne Member of the Diest Formation revisited: Unexpected results from the study of a large temporary outcrop near Antwerp International Airport, Belgium. *Geol. Belg.* **2020**, *23*, 219–252. [CrossRef]
169. Villafaña, J.A.; Kindlimann, R.; Chavez-Hoffmeister, M. The confirmed fossil record of the blue shark *Prionace glauca* (Linnaeus, 1758) from the South Eastern Pacific. *Span. J. Palaeontol.* **2022**, *37*, 249–256. [CrossRef]
170. Bellocchio, G.; Carboni, M.G.; Nami, M.; Pallini, G. Fauna ad ittiodontoliti del Pliocene di Allegrona (Terni, Umbria). *Boll. Soc. Nat. Napoli* **1991**, *100*, 41–73.
171. Cook, T.D.; Murray, A.M.; Simons, E.L.; Attia, Y.S.; Chatrath, P. A Miocene selachian fauna from Moghra, Egypt. *Hist. Biol.* **2014**, *22*, 78–87. [CrossRef]
172. Bianucci, G.; Di Celma, C.; Collareta, A.; Landini, W.; Post, K.; Tinelli, C.; de Muizon, C.; Bosio, G.; Gariboldi, K.; Gioncada, A.; et al. Fossil marine vertebrates of Cerro Los Quesos: Distribution of cetaceans, seals, crocodiles, seabirds, sharks, and bony fish in a late Miocene locality of the Pisco Basin, Peru. *J. Maps* **2016**, *12*, 1037–1046. [CrossRef]
173. Marsili, S. Revision of the teeth of the genus *Carcharhinus* (Elasmobranchii; Carcharhinidae) from the Pliocene of Tuscany, Italy. *Riv. Ital. Paleontol. Stratigr.* **2007**, *113*, 79–96.
174. Betancort, J.F.; Lomoschitz, A.; Meco, J. Early Pliocene fishes (Chondrichthyes, Osteichthyes) from Gran Canaria and Fuerteventura (Canary Islands, Spain) [Los peces (Chondrichthyes, Osteichthyes) del Plioceno inferior de Gran Canaria y Fuerteventura (Islas Canarias, España)]. *Estud. Geológicos* **2016**, *72*, e054. [CrossRef]
175. Govender, R.; Chinsamy, A. Early Pliocene (5 Ma) Shark-Cetacean Trophic Interaction from Langebaanweg, Western Coast of South Africa. *Palaios* **2013**, *28*, 270–277. [CrossRef]
176. Purdy, R.W. The early Miocene fish fauna from the Pollack Farm Site, Delaware. In *Geology and Paleontology of the Lower Miocene Pollack Farm Site, Delaware*; Benson, R.N., Ed.; Delaware Geological Survay, Special Publication: Adelaide, Australia, 1998; Volume 21, pp. 133–139.
177. Collareta, A.; Merella, M.; Casati, S.; Di Cencio, A. First fossils of the extant blacktip shark *Carcharhinus limbatus* from Europe and the Mediterranean Basin. *Neues Jahrb. Geol. Paläontologie Abh.* **2021**, *301*, 109–118. [CrossRef]
178. Sahni, A.; Mehrotra, D.K. The elasmobranch fauna of coastal Miocene sediments of peninsular India. *Biol. Mem.* **1981**, *5*, 83–121.
179. Sharma, K.M.; Patnaik, R. Miocene fishes from Baripada Bbeds, Orissa and their palaeoenvironmental, palaeobiogeographic and palaeoclimatic significance. *Spec. Publ. Palaeontol. Soc. India* **2014**, *5*, 291–323.
180. Mora Morote, P. Peces Galeomorfos y Squatinomorfos en el Pliocène de Guardamar del Segura (Alicante). *Cidaris Rev. Ilicitana Paleontol. Mineral.* **1996**, *5*, 98–124.
181. Cappetta, H. *Handbook of Paleoichthyology, Vol. 3B: Chondrichthyes II*; Gustav Fischer Verlag: Stuttgart, Germany, 1987.
182. Compagno, L.J.V. *Sharks of the Order Carcharhiniformes*; Princeton University Press: Princeton, NJ, USA, 1988.
183. Costa, S.A.F.; Richter, M.; Toledo, P.M.; Moraes-Santos, H.M. Shark teeth from Pirabas Formation (Lower Miocene), northeastern Amazonia, Brazil. *Bol. Mus. Emilio Goeldi Ciências Nat.* **2009**, *4*, 221–230. [CrossRef]

184. Landini, W.; Altamirano-Sierra, A.; Collareta, A.; Di Celma, C.; Urbina, M.; Bianucci, G. The late Miocene elasmobranch assemblage from Cerro Colorado (Pisco Formation, Peru). *J. S. Am. Earth Sci.* **2017**, *73*, 168–190. [CrossRef]
185. Applegate, S.P. The El Cien Formation, Strata of Oligocene and early Miocene Age in Baja California Sur. *Univ. Nac. Auton. Mex. Inst. Geol. Rev.* **1986**, *6*, 145–162.
186. Sánchez-Villagra, M.R.; Burnham, R.J.; Campbell, D.C.; Feldmann, R.M.; Gaffney, E.S.; Kay, R.F.; Lozsán, R.; Purdy, R.W.; Thewissen, J.G.M. A new near-shore marine fauna and flora from the early Neogene of northwestern Venezuela. *J. Paleontol.* **2000**, *74*, 957–968. [CrossRef]
187. Iturralde-Vinent, M.A.; Hubbell, G.; Rojas, R. Catalogue of Cuban fossil Elasmobranchii (Paleocene to Pliocene) and paleogeographic implications of their lower to middle Miocene occurrence. *J. Geol. Soc. Jam.* **1996**, *31*, 7–21.
188. Portell, R.W.; Hubbell, G.; Donovan, S.K.; Green, J.L.; Harper, D.A.T.; Pickerill, R. Miocene sharks in the Kendeace and Grand Bay formations of Carriacou, The Grenadines, Lesser Antilles. *Caribb. J. Sci.* **2008**, *44*, 279–286. [CrossRef]
189. Aguilera, O.; Luz, Z.; Carrillo-Briceño, J.D.; Kocsis, L.; Vennemann, T.W.; de Toledo, P.M.; Nogueira, A.; Amorim, K.N.; Moraes-Santos, H.; Polck, M.R.; et al. Neogene sharks and rays from the Brazilian 'Blue Amazon'. *PLoS ONE* **2017**, *12*, e0182740. [CrossRef]
190. Carrillo-Briceño, J.D.; Aguilera, O.A.; de Gracia, C.; Aguirre-Fernández, G.; Kindlimann, R.; Sánchez-Villagra, M.R. An Early Neogene Elasmobranch fauna from the southern Caribbean (Western Venezuela). *Palaeontol. Electron.* **2016**, *19.2.27A*, 1–32. [CrossRef]
191. Caretto, P.G. Osservazioni tassonomiche su alcuni Galeoidei del Miocene piemontese. *Boll. Soc. Paleontol. Ital.* **1972**, *11*, 14–85.
192. Fialho, P.R.; Balbino, A.C.; Antunes, M.T. Fossil Chondrichthyes from the Neogene of Portugal: Diversity and Occurrence. *Anuário Inst. Geociências* **2021**, *44*, 43395. [CrossRef]
193. de Stefano, G. Osservazione sulle ittiofauna pliocenica di Orciano e San Quirico in Toscana. *Boll. Soc. Geol. Ital.* **1909**, *28*, 539–648.
194. Macphee, R.D.E.; Iturralde-Vinent, M.A.; Gaffney, E.S. Domo de Zaza, an Early Miocene Vertebrate Locality in South-Central Cuba, with Notes on the Tectonic Evolution of Puerto Rico and the Mona Passage. *Am. Mus. Novit.* **2003**, *3394*, 1–42. [CrossRef]
195. Macfadden, B.J.; Jones, D.S.; Jud, N.A.; Moreno-Bernal, J.W.; Morgan, G.S.; Portell, R.W.; Perez, V.J.; Moran, S.M.; Wood, A.R. Integrated Chronology, Flora and Faunas, and Paleoecology of the Alajuela Formation, Late Miocene of Panama. *PLoS ONE* **2017**, *12*, e0170300. [CrossRef] [PubMed]
196. Villafaña, J.A.; Rivadeneira, M.M.; Pimiento, C.; Kriwet, J. Diversification trajectories and paleobiogeography of Neogene chondrichthyans from Europe. *Paleobiology* **2023**, *49*, 329–341. [CrossRef]
197. Höltke, O.; Maxwell, E.E.; Pollerspöck, J.; Rasser, M.W. The shark and ray teeth of the Lower Miocene (Upper Marine Molasse) from Äpfingen, Baden-Württemberg, Southern Germany. *Neues Jahrb. Geol. Paläontologie Abh.* **2022**, *305*, 323–342. [CrossRef]
198. Stein, R.W.; Mull, C.G.; Kuhn, T.S.; Aschliman, N.C.; Davidson, L.N.; Joy, J.B.; Smith, G.J.; Dulvy, N.K.; Mooers, A.O. Global priorities for conserving the evolutionary history of sharks, rays and chimaeras. *Nat. Ecol. Evol.* **2018**, *2*, 288–298. [CrossRef] [PubMed]

diversity

MDPI

Article

First Use of Free-Diving Photo-Identification of Porbeagle Shark (*Lamna nasus*) off the Brittany Coast, France

Armelle Jung *, Arthur Ory, Paul Abaut and Lucas Zaccagnini

Des Requins et Des Hommes (DRDH), Bibliothèque la Pérouse Technopole Brest Iroise, 15 Rue Dumont d'Urville, BP 70F, 29280 Plouzané, France; arthur.ory@desrequinsetdeshommes.org (A.O.); paul.abaut@desrequinsetdeshommes.org (P.A.); lucas.zaccagnini@desrequinsetdeshommes.org (L.Z.)

* Correspondence: armelle.jung@desrequinsetdeshommes.org; Tel.: +33-(0)614386001

Abstract: A large number of pelagic shark species have declined significantly in recent decades due to overfishing, bycatch, and habitat degradation. Whereas porbeagle sharks have become scarce due to a reduction in their populations around the world, recent stock evaluations are giving positive signals about the evolution of the North-Eastern Atlantic stock size. The porbeagle shark (*Lamna nasus*), an offshore pelagic species with a wide distribution, is designated by IUCN as Globally Vulnerable and Critically Endangered for Europe and subject to various international conservation conventions. An increasing number of observations are reported off the Brittany coast of Trégor. The ecological role of this area for the species is still unknown and greater knowledge is needed to develop and apply sustainable management measures on a local and international scale. This study represents the first use of photo-identification on porbeagle sharks in order to improve the ecological knowledge of the species in the Trégor area. These results confirm the effectiveness of this method, with 19 of the 131 individuals identified being re-sighted, indicating an interesting degree of site fidelity and showing a sex ratio of 100% females. Observations of individuals over several years allowed the researchers to discuss the relevance of the different types of marks. The findings suggest that the Trégor area off the Brittany coast serves as a seasonal residence for female porbeagle sharks, especially between May and October. This study represents a successful first step in the use of photo-identification for this species. It offers technical support for the sharing of the methodology and provides some biological knowledge allowing researchers to discuss potential sustainable management measures for the conservation of porbeagle sharks in the study area and their habitats while needed.

Keywords: *Lamna nasus*; photo-identification; free-diving; site fidelity; ecology

1. Introduction

Porbeagle shark *Lamna nasus* (Bonnaterre, 1788) is a pelagic and neritic species widely distributed among the Northern and Southern Atlantic, the Mediterranean Sea, and the southern Indian and Pacific Oceans [1]. Porbeagle sharks can reach a large body size and perform long seasonal movements [2,3]. *Lamna nasus* mainly feeds on small pelagic fishes and cephalopods [4] although its feeding behavior is considered opportunistic because its diet varies among regions [5,6]. Since the 1930s, porbeagle sharks were subject to commercial fisheries that resulted in a severe decrease in their North Atlantic populations [7,8]. Listed in 2006 as Globally Vulnerable [9] and Critically Endangered for Europe [10] by the IUCN Red list, the porbeagle shark was also included in various international conventions such as the UN Convention on the Law of the Sea (UNCLOS) Annex I, the Convention International Trade of Endangered Species (CITES) Appendix II, the Convention of Migratory Species (CMS) Appendix II, the Barcelona Convention Annex II, and the Bern Convention Appendix II, highlighting its degree of vulnerability and the need of cooperative management. Porbeagle sharks were subject to fishery regulation by the Council of the European Union in

2010 prohibiting EU vessels in Union waters and in certain non-Union waters to fish for, to retain on board, to transship, and to land these sharks in International European waters [11].

The use of photography to discriminate animals has been broadly used to study large terrestrial species [12–14]. These techniques of photo-identification (Photo-ID) spread in marine research mainly for marine mammals [15–17] or elasmobranch species able to be encountered at the surface such as whale shark (*Rhincodon typus*), basking shark (*Cethorinus maximus*), spotted eagle (*Aetobatus narinari*), or manta rays (*Manta alfredi*) [18–22]. More recently, the use of photo-ID has become much more popular among shark scientists [23]. While extending the species coverage, different parts of the shark body were considered: spots next to the gills for leopard sharks (*Stegostoma fasciatum*) [24], the shape of their fins for white sharks (*Carcharodon carcharias*) [25], basking shark (*Cethorinus maximus*) [20], and nurse shark (*Ginglymostoma cirratum*) [26], fins and body scars for the nurse shark (*Ginglymostoma cirratum*) [27] and basking shark (*Cethorinus maximus*) [20], or the pattern of their body stripes for species such as tiger sharks (*Galeocerdo cuvier*) [28]. These capture-recapture approaches have been described as tools to estimate the abundance of marine megafauna species [29–31]. Whereas porbeagle sharks have become scarce due to a reduction in their populations around the world, recent stock evaluations are giving positive signals about the evolution of the North-Eastern Atlantic stock size [32]. The seasonal presence of porbeagle shark off the Brittany coast of Trégor has been documented since the 1960s [33]; however, the recent increase in sightings reported by locals aroused the interest of our team.

The study area is the Trégor region, located on the northern coast of Brittany (France), in the Côtes d'Armor territory (Figure 1). This area has a particular geological and bathymetric structure, with a steep rocky slope very close to the coast. Trégor is recognized for its marine biodiversity richness which supported the implementation of the marine protected area called National Nature Reserve of Septs Iles in 1976 [34–36], also classified under European Habitat and Bird Directives. This area, also named Pink Granite coast, features a specific magnetic anomaly, linked to the origins of its geological formation [37]. As sharks are very sensitive to electromagnetic fields [38–40], the regular presence of porbeagle sharks so close to the coast could be linked to the presence of this magnetic anomaly.

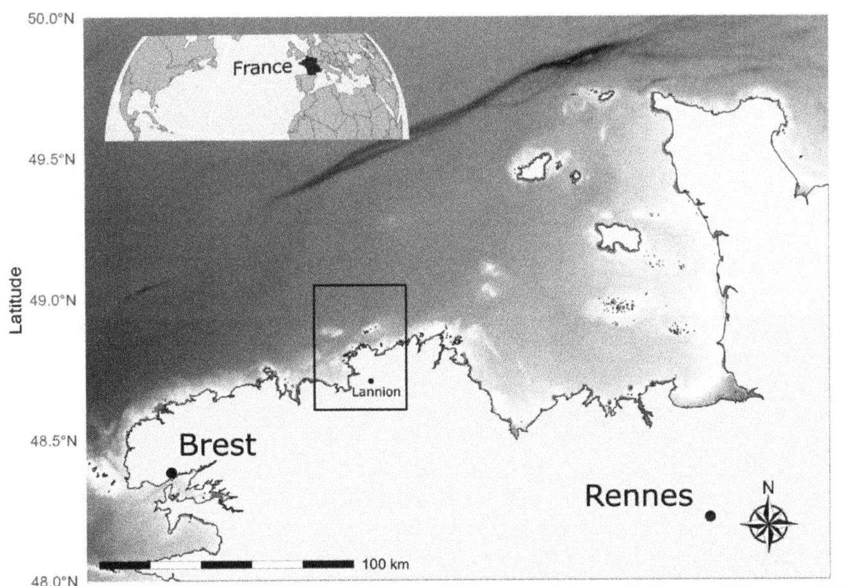

Figure 1. Study area of the porbeagle photo-ID program, DRDH (2021–2023).

In 2021, we started a free-diving protocol based on the methodological recommendations made by Marshall and Pierce [23] for such a photo-ID approach in order to minimize methodological biases. The objectives are (1) to evaluate the presence of porbeagle sharks in the area, (2) to develop an individual shark catalog and assess which phenotypic criteria are most relevant, and (3) to improve the ecological knowledge of the species.

As suggested by the isotopic analysis conducted on individuals coming [41], some degree of intra-population heterogeneity exists between inshore and offshore individuals, and would demand further investigation on their movement behavior. This research allows us to implement a precursory approach to study the use of this very coastal zone for the species and offers a reproducible and collaborative tool to share among research groups working on porbeagles.

2. Materials and Methods

2.1. Field Sampling and Underwater Observations

From June 2021 to September 2023, the DRDH (Des Requins et Des Hommes) team conducted 64 free-diving sessions (corresponding to an estimated 110 h of immersion) following the same protocol. The vessels used to reach the survey area were semi-rigid boats from the professional diving center Joly Plongée (10 m, 250 HP) or DRDH's own boat (Zodiac 5.5 m, 80 HP). For each dive, a group of two to six trained divers used a lead drifting line of 15 m long (stainless steel) equipped with a nylon bag containing 500–1000 g of bait as olfactive attraction only. Bait included sardine (*Sardina pilchardus*) or mackerel (*Scomber scombrus*), fresh or frozen according to the fish market availability. The divers stayed in pairs at the surface and made regular dive checks up to 10 m depth until a shark approached. In order to adhere to our low-intrusive code of conduct, the dive time was limited to 1 h 30 min, and the dive was finalized with or without shark encounter after this period of research.

In the case of an animal coming by, the free divers quietly went down up to 10 m and waited in apnea for the shark to approach in order to film the interaction. Underwater observations were video-recorded with underwater cameras Canon G7x mk2 (Canon, Tokyo, Japan) or GoPro Hero+, 7, 9, 10, and 11 (GoPro, San Mateo, CA, USA). Since 2022, the cameras have been equipped with a mounted pair of underwater lasers (Green Laser 18650 Li-ion 3.6 V, Oceanco Ltd., FL, USA) of 30 cm distance to measure the length of a shark. Before 2022, or when laser measuring was not possible, size estimations were based on in situ observation or videos post-evaluation by comparison to diver size.

2.2. Data from Maritime Stakeholders

In order to increase the geographical and temporal coverage of the shark sightings data, some additional footage taken by professional photographers was collected and included in the photo-ID catalog. Social networks such as Facebook, YouTube, and some dedicated internet forums (spearfishing, recreational fishing, or scuba diving websites) were visited to check some publications of shark sightings. In case the quality of the picture and the detail of information was adequate, the footage owner was contacted to request their permission for the research use of the picture or video. These external data were only used as qualitative data to improve the photo-ID catalog and will not be included in population size analyses or any other analysis requiring a standardized sampling protocol and effort.

2.3. Phenotypic Markers

The identification criteria chosen for the porbeagle shark individual discrimination were as follows: (1) the sex, by checking the presence of the external male reproductive organ; claspers are visible next to the pelvic fin; (2) the size, calculated via laser photometry or via visual estimation; and (3) the body marks (described in Figure 2) as well as the shape and color of the fins and countershading delineation.

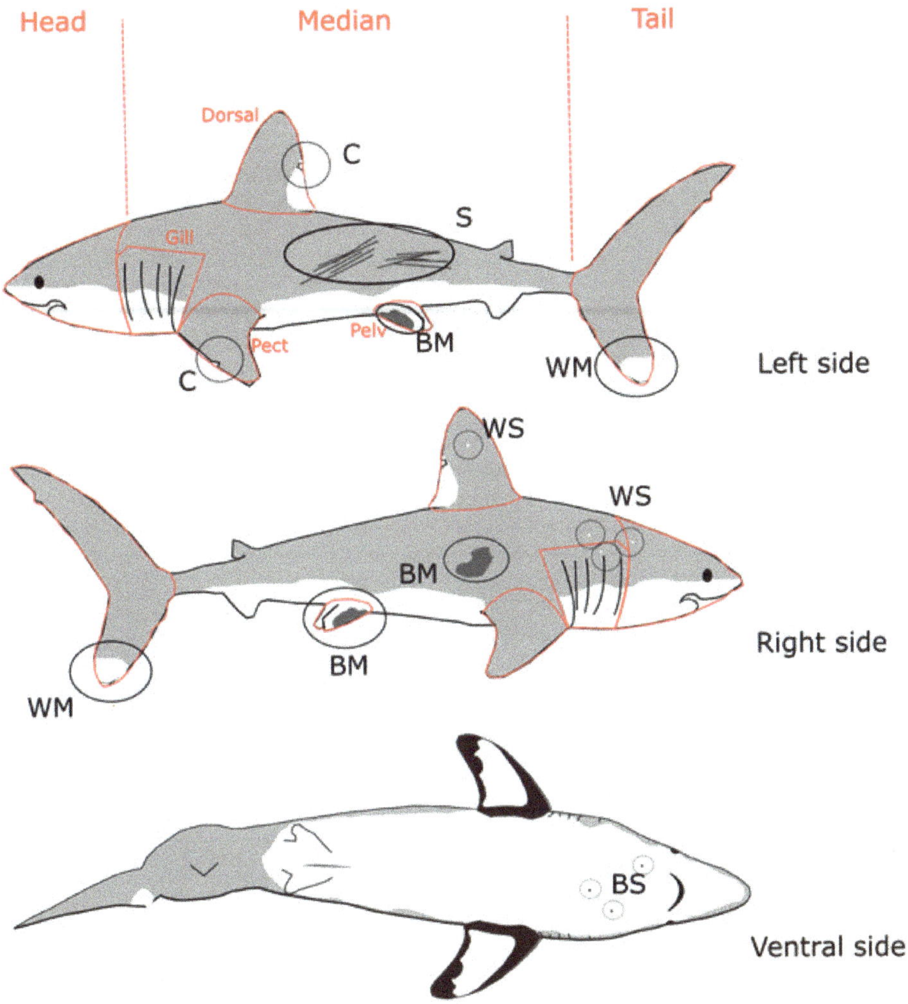

Figure 2. Body parts and marks used for porbeagle shark individual discrimination: Tail, Median, Gill, Head, Pelvic Fin, Pectoral Fin, Dorsal Fin. S = Scar, BM = Black Mark, C = Clip, WM = White Mark, WS = White Spot, BS = Black Spots, presented on the Left, Right, or Ventral sides.

These marks are used in many other shark photo-identification studies, especially for tiger sharks (*Galeocerdo cuvier*) [28], great white sharks (*Carcharodon carcharias*) [42,43], and basking sharks (*Cethorinus maximus*) [20].

2.4. Additional Data

In order to improve the ecological knowledge of the species, other parameters were collected during the fieldwork for future analyses and were therefore not used in this study. We used a hand sensor to collect sea surface temperature (SST) and salinity (Mettler Toledo, Greifensee, Suisse). Additional environmental parameters such as tide coefficient, cloud cover, wave height, moon phase, and the number of divers were also collected for each dive. We also checked and documented the presence of macro parasites such as copepods while checking the photos. Some additional elements were also recorded such as anthropogenic remains: net entangle cicatrix, hook lines hanging, tagging devices, relics, or evidence of interaction with boats (e.g., propeller scars).

2.5. Data Treatment and Analysis

To distinguish the animals, the videos were meticulously analyzed via VLC player (OpenSource, France) or Movie & TV (Microsoft, WA, USA), especially using the slow-motion facility. Screenshots of each side of the shark and details of specific criteria were taken and stored in an individual folder by date. To facilitate the data analysis, the body was divided into seven different parts (Head, Median, Tail, Gill, Dorsal Fin, Pectoral Fin, and Pelvic Fin) for each side of the shark and assigned to six types of marks (Scar (S), Black Mark (BM), Clip (C), White Mark (WM), White Spot (WS), and Black Spot (BS)) as described in Figure 2. Retained criteria were entered both in the Excel folder catalog and in a corresponding single ID paper form; a single ID-code was allocated to each shark—for practical reasons, an additional nickname was also used for animals we used to encounter. The maturity stage of the females was estimated based on the maturity size observed by Jensen et al. [44] in the North Atlantic. As no males were observed during this study, no visual method for estimating the maturity stage of males was developed. Because this study is the first to use photo identification on this species, all the marks were considered at the same level. When a new observation was included in the database, a simple R function was used to assist potential assignment to an individual which had already been observed. Basically, the function searches in our database for all observations corresponding to individuals showing a minimum number of similar common criteria (the same type of mark on the same area and side of the body). This number of criteria is chosen according to the number of available visible criteria of the new observation and adjusted in order to have up to 20 potential matches found by the function. All the potential matches are then compared visually with the new observation to check if the marks are considered the same or not, allowing the new observation to be assigned or not to an individual already recorded in the database. If the observation is not assigned to an individual already registered, a new individual code is created in the database. To avoid reading discrepancies, all assignments were made by the same observer; in cases of uncertainty, other members of the team were consulted for double reading.

3. Results

3.1. Porbeagle Photo-ID Catalog

The porbeagle photo-ID program, in three years of field seasons, was successful in making 183 shark sightings which correspond to 131 different individuals (100% females) ranging from 1.5 m to 2.5 m fork length (Lf) observed between 2015 and 2023 in the study area. The majority of the sightings came from DRDH diving sessions (60%), contributions from professional photographers represented 28%, and underwater online publications 12%. No particular trend is visible in the time series so far; from the three years of free-diving seasons performed by DRDH, 2023 is the year with the most porbeagle sighting events whereas external sighting data are the most important for 2020 (Figure 3).

Some obvious phenotypic marks such as large fin clips or massive body scars allowed us to easily recognize some of the sharks. However, in most cases, identification was possible due to a combination of morphological marks associated with body size. All assignments were made using a minimum of two and a maximum of eleven criteria (an average of six criteria). As illustrated for the shark n°20LAMNA#13 ("Mylene"), the white marks (WM) discerned on the dorsal fin are well recognizable (Figure 4A), as highlighted by the red circles drawn as post-treatment. Three white spots (WS) visible on the left side upper gill are also clear over time (Figure 4A) although they require careful screenshot analysis and good light exposure. The white mark (WM) present on the lower caudal fin is also very characteristic and constitutes a good criterium to recognize "Mylene", even when in the water (Figure 4C). Finally, the observation of the countershading delineation (Figure 4D) is very efficient for photo-recapture criteria because this pattern is unique for each individual and persists for several years (Figure 4D).

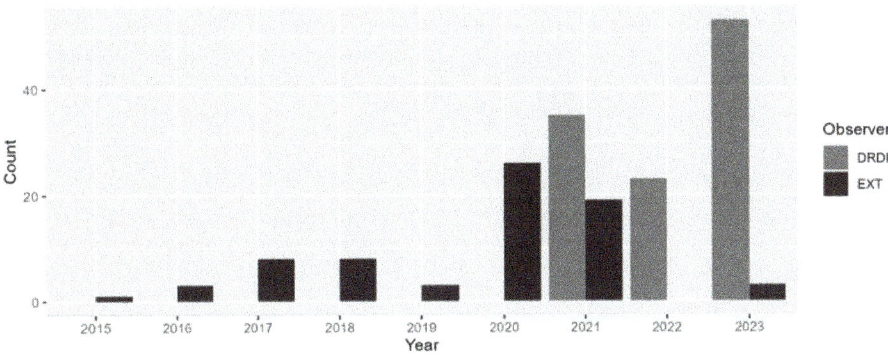

Figure 3. Total number of porbeagle shark sightings between 2015 and 2023 by DRDH and other marine stakeholders (EXT).

Figure 4. Example of the major persistent body marks for porbeagle sharks for individual n°20LAMNA#13 ("Mylene"): (**A**) Dorsal fin and upper gills, (**B**) general body view from left side, (**C**) tail, and (**D**) countershading delineation from 2020 to 2023. Red dashed lines are representing the color contrasts delineations post-added manually.

3.2. Seasonal Variations

Sightings occurred between March and September. For the dives performed by DRDH, no significant difference can be observed within the months while weighed by the effort of research ($p > 0.05$, Tukey test). However, July is the month showing the highest variability in the number of animals observed per dive (cf. Figure A1, additional material).

3.3. Re-Sighting Rate and Site Fidelity

Over the 131 individual sharks identified, from 2015 to 2023, the photo-ID techniques allowed 19 different animal re-sightings corresponding to 51 re-sighting events (Figure 5). Re-sightings occurred from 1 to 10 times maximum for the shark n°20LAMNA#13 ("Mylene"), 57% of re-sighted sharks were observed twice, 15% were observed three times, and 25% were observed more than four times. Most of the sharks were photo-recaptured during the same year (68%), with the longest time series of four consecutive years for four animals. The longest period between two re-sightings was four years, and this event occurred once for a mature female shark n°17LAMNA#07 between 2017 and 2021. We observed a maximum of five re-sightings during the same season in 2023 for "Yvette", the porbeagle shark n°20LAMNA#18. The large majority of the re-sightings involved mature porbeagle sharks (82% of the sightings events and 58% of the number of individuals).

Shark Code	Nick Name	Estimated Size LF (m)	Sex	Maturity stage	Year of observation									Total Nb of resighting
					2015	2016	2017	2018	2019	2020	2021	2022	2023	
20LAMNA#13	Mylene	2.3	F	Mature						2	3	2	4	10
20LAMNA#17	Fanny	2.3	F	Mature						2	3	1*	4*	9
20LAMNA#18	Yvette	2.3	F	Mature						1	2	1	5	8
21LAMNA#03	Armelita	2.3	F	Mature							1*	2	3	5
23LAMNA#19	Allana	2.1	F	Mature									4	3
21LAMNA#08	Elisabeth	2.0	F	Sub-adult							3			2
21LAMNA#10	Monique	2.0	F	Mature							2		1	2
21LAMNA#15	Nikita	1.9	F	Sub-adult							3			2
15LAMNA#01	Jumpa	2.1	F	Sub-adult	1	1								1
17LAMNA#07	-	2.0	F	Mature			1				1			1
18LAMNA#01	-	2.0	F	Mature				1		1*				1
18LAMNA#03	-	2.1	F	Mature				2						1
19LAMNA#03	-	2.2	F	Mature					1		1			1
20LAMNA#05	Didy	1.9	F	Sub-adult						1	1			1
21LAMNA#06	-	1.7	F	Juvenile							2			1
21LAMNA#33	Anne	2.2	F	Mature							2			1
22LAMNA#15	-	1.8	F	Juvenile								2		1
23LAMNA#08	-	1.8	F	Sub-adult									2	1
23LAMNA#15	-	1.8	F	Juvenile									2	1
Total														**51**

Figure 5. Total number of re-sighting events for porbeagle sharks observed by the DRDH team and marine stakeholders between 2015 and 2023. Maturity scale ($\leq$1.8 m = Juvenile, 1.8–2.2 m = Sub-Adult, $\geq$2.2 m = Mature—all measures are fork length); (*) Gravid individuals.

3.4. Biological Information

Underwater observation allows us to document some elements on the maturity of the sharks visiting the area (Figure 5). We encountered four gravid females including one n°20LAMNA#17 ("Fanny") twice in consecutive years: in June 2022 and June 2023. The same individual was observed in the postpartum stage in August 2023, with an evident reduction in the belly volume between the two final observations (cf. Figure A2, additional material).

While scrutinizing the Photo-ID catalog, we realized that some phenotypic criteria were less constant than others over time, for the porbeagle sharks we encountered; for instance, scars tend to reduce in size over time. For the porbeagle shark n°21LAMNA#15

("Nikita"), we estimated a reduction of 75% in the scar surface within 62 days, as presented in Figure 5.

Moreover, we noticed an interesting phenomenon for the porbeagle shark "Mylene": black spots (BS) were present around the gills on both sides of this animal during the encounter that occurred in July 2021, but they were not visible one year before (Figure 6). When "Mylene" was photo-recaptured two months later in August 2021, these BS were no longer present.

Figure 6. Example of the non-persistence of certain scars (S) for the porbeagle shark n°21LAMNA#1.5 ("Nikita").

4. Discussion

4.1. Effectiveness of the Methodology

This study shows for the first time the feasibility and efficiency of using photo identification to distinguish and study individual porbeagle sharks over several years. The re-sighting rate of porbeagle sharks in this study is 15%. As this is the first study using photo identification on this species, comparison with other sites or populations is not yet possible. However, comparable photo-ID methodologies with similar temporal coverage have shown highly variable re-sighting rates among species. For instance, a re-sighting rate of only 5% has been observed for basking sharks [20], whereas such rate is 61% for tiger sharks [28] and 47% for nurse sharks [27]. The re-sighting rate can also vary depending on the study area. For the great white shark, Domeier and Nasby Lucas [42] observed a re-sighting rate of 78% at Guadalupe Island in Mexico, while Hewitt et al. [25] observed a rate of 29% at Seal Island in South Africa. Finally, re-sighting rates are also linked to the methodology of observation (e.g., at sea survey, BRUV, citizen sciences, or drone) but also depend on the ecology of the studied species. Considering the scarcity of porbeagle sightings underwater, the re-sighting rate in this study can be considered satisfactory and adequate to continue the implementation of the methodology. As the photo-identification protocol has only been used for three years and the sampling effort has increased, it is expected that this rate will increase in the coming years.

The variety of distinctive marks as well as the multiple re-sightings of some individuals will make it possible to determine which types of marks are the most relevant for long-term identification. Countershading delineation, black marks on the belly and pelvic fins, white marks on the caudal and dorsal fins, and white spots seem practical for long-term identification (Figures 2 and 3). However, countershading delineation requires time-consuming post-treatment and good exposure in terms of light and orientation of the body; for instance, the screenshot available for 30 August 2021 (Figure 4D) does not allow the use of this phenotypic mark. It is therefore not always possible to use this mark to confirm an identification. These kinds of marks are also used for the long-term identification of tiger sharks (*Galeocerdo cuvier*) [28] and great white sharks (*Carcharodon carcharias*) [42]. In this study, scars on the body and the gills and fin clips were used to identify individuals in the short term. But it is important to be aware that these marks can change over time, with small scars reducing over the years or being replaced by a new, larger scar, leading to serious problems for long-term identification (Figure 6). This has already been observed for great white sharks (*Carcharodon carcharias*) [43], tiger sharks (*Galeocerdo cuvier*) [28], and basking sharks (*Cethorinus maximus*) [20]. However, the use of multiple marks can increase the accuracy of individual identification, thereby reducing observer bias and the probability of false identification due to changes in some marks [28,42,45]. Graham and Roberts [21] even considered that photo identification based on multiple marks is more efficient than traditional visual tagging methods. Indeed, despite a significant change in the general appearance of some of the sharks identified (Figures 6 and 7), the use of several markings enabled them to be properly identified. In addition, this study observed for the first time changes in darker pigmentation spots on porbeagle sharks over a short period of time (Figure 7). Domeier and Nasby-Lucas [42] have already observed a similar phenomenon on the great white shark (*Carcharodon carcharias*). Black spots are thus not used to identify individuals in this study as standalone, but combined with other criteria.

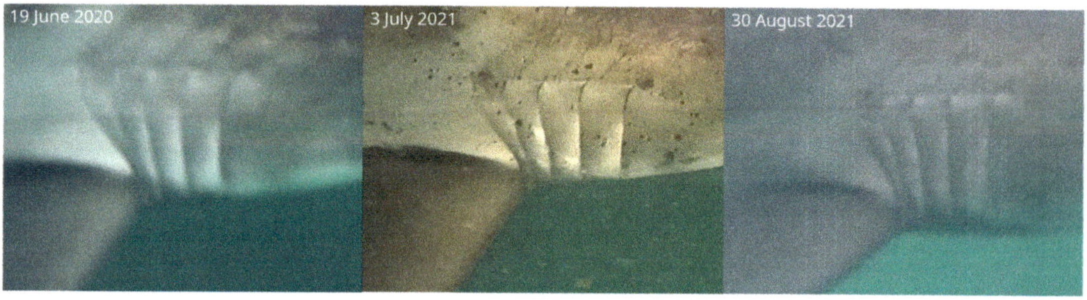

Figure 7. Pigmentation variations observed on the same porbeagle shark n°20LAMNA#13 ("Mylene") observed in June 2020, July, and August 2021.

Furthermore, this underwater photo-ID approach can be considered as an asset in comparison to traditional recapture methodology such as tagging protocols. Free diving allows approaching the shark in a silent and smooth way, avoiding disturbance and reducing potential behavioral modification (attraction or avoidance) although a long-term study on shark behavior is required. Moreover, photo-ID is considered as a less impactful method by preventing capture or shark manipulation that affect animal fitness through metabolic stress [23] or potential epizootic events.

4.2. Ecology

The seasonality observed with the first results of this study is consistent with current knowledge. Porbeagle shark landing data from the French targeted fishery [46] indicate a strong presence of porbeagle sharks in spring and summer in shallow waters on the north-east Atlantic continental shelf, and tagging studies suggest autumn and winter migration behavior in offshore deep water [2,47]. [7] observed the same pattern along the Canadian coast, based on data from inshore and offshore fisheries.

The life cycle of the species is also an element that this methodology has allowed us to document. Porbeagle shark n°20LAMNA#17 ("Fanny") was observed gravid twice, within a 12-month interval. Current knowledge estimates the gestation period for porbeagle sharks around 8 to 9 months probably on an annual or potentially biennial period [44,48], which suggests that this female gave birth twice in two consecutive years. Furthermore, the observations suggest that "Fanny"'s parturition occurred between June and August 2023, potentially in the study area. Although additional research is required to state the ecological functionality of this area for the species, the overall findings suggest that the Tregor area, and the seven islands MPA in particular, could play a major role in the life cycle of this group of porbeagle. This non-invasive biological information, which is difficult to obtain using traditional methods, such as scientific fisheries, electronic tagging, or ultrasound technologies [49,50], can benefit the global understanding of the species' ecology.

It is well established that shark species can exhibit sex-specific segregation and resident areas, and their migration patterns can be dependent on their sex [51–53]. As mentioned above, a wide range of porbeagle shark sizes is observed in this study (1.5 m to 2.5 m LF) and it appears that this particular area serves as a seasonal aggregation zone where exclusively females were able to be detected. This finding is consistent with the study of Hulbert et al. [54] on salmon sharks (*Lamna ditropis*) which reports a concentration of females (95% sex ratio) with a similar size range in Prince William Sound (Alaska). Similar findings have been described on blue sharks (*Prionace glauca*) by Druon et al. [55], who observed that large females' habitats tend to overlap with juveniles. Nevertheless, further studies are needed to assess potential variations in ecological behavior based on size or sex, such as whether males or juveniles exhibit deeper, offshore habitats with reduced coastal or exploratory movements, or simply show less interest in the "scent trail". Hennache and Jung [56] have documented variation in the sex ratio of porbeagle sharks in the Bay of Biscay based on fishery-dependent information, ranging from 1 female for every 0.74 males in the South Irish region to 1 female for every 1.19 males in the Canal St. Georges. But until now, no other area occupied exclusively by porbeagle shark females had been identified in the Northeast Atlantic. This situation has already been observed for the great white shark (*Carcharodon carcharias*), as published for Southern Australia by Bradford et al. [57], where only females are encountered in winter time. The existence of these sex-specific aggregation sites is increasingly well identified, but remains difficult to interpret.

4.3. Caveats and Limitations

Underwater methodologies are known to present some limits such as the visibility or sea conditions; this parameter can be of particular concern in Brittany waters (high tides, strong current, and low temperatures). This can affect both the detection of the animal and the possibility to acquire good quality footage. Given that the bottom depth in the study area was mainly around 40 m, with visibility varying from 2 to 15 m, some porbeagle sharks may have visited our team without possible detection. Additionally, the performance of the methodology is reduced in the case of high turbidity (in the case of plankton blooms, for instance).

Furthermore, a correct approach of the animal is crucial to obtain good footage and measurements. It requires the training of good free divers, accustomed to difficult sea conditions and able to interpret and predict shark movement in order to be in the right position while the animal will pass by them.

Nevertheless, with more than 130 individuals identified, the database has reached a considerable size, resulting in a significant increase in the time required to assign an observation to an individual, already recorded or not. The use of automatic or semi-automatic identification methods such as collaborative multi-user software (Wildbook for shark, Microsoft, WA, USA) or the mapping of pigmentation patterns with specific software such as "I3S" [26,58] could be a relevant solution. However, these methods require a standard identification protocol and good-quality images taken from the right angle [45,59]. Conditions of high turbidity and low visibility, as well as the use of small sports cameras, could reduce the efficiency of these methods for this study. Efficiency tests are therefore required before applying these methods in future years of monitoring.

4.4. Perspectives

Additional years of photo-identification monitoring coupled with other methods are needed to better understand the porbeagle shark's use of this area. The validation of the photo-identification method by genetic identification, the estimation of the size of the group present in the area, and the study of family relationships between individuals observed [60] would provide a better understanding of the importance of this area in the porbeagle shark's life cycle and ecology. The analysis of the relationships between the various environmental factors and the number of shark sightings will also be carried out once more data have been collected to ensure the accuracy of the subsequent modeling process. Additionally, systematic laser measurements will be conducted in the future and will enable a more comprehensive description of the size distribution as well as the growth rate estimation for re-sighted animals.

Moreover, the individual underwater identification may complement the ecological knowledge of the species as it allows the collection of multiple additional samples such as the experiment in summer 2023 with skin biopsy and eDNA water filtration. In the meantime, the team dives for individual-specific recordings, which are key elements for research on the group structure and connectivity [50,61].

Finally, estimating anthropic pressures on a local population is complementary to understanding ecology [62], and relevant to implementing appropriate local management adapted to the area. Because the study area is on the border of an MPA (Marine Protected Area), the recent touristic interest for this species and the increase in human interactions, particularly bycatch and depredation (personal communication C. Mangin 17 July 2023—Recreational fishery committee president), also highlight the need for a thorough study of the behavior and ecology of the porbeagle shark in this area, in order to adopt concerted management methods. In addition, the significant proportion of our database obtained from local stakeholders (40%) highlights the importance and relevance of citizen science for the study of marine predators ranging from sharks [63] to marine mammals [64].

5. Conclusions

This study enabled us to establish the first photo-identification catalog for porbeagle sharks. The method proved effective, as 19 of the 131 females identified were re-sighted, indicating a certain degree of site fidelity. Given that sightings are most frequent between May and October, the Trégor area seems to serve as a seasonal residence for female porbeagle sharks, although the specific ecological importance of this site remains to be determined. It should be noted that males may also be present at the site, but their presence may not have been detected. Future research using alternative methods will be essential to better understand porbeagle shark ecology in this area and beyond, enabling the implementation of management measures adapted to the well-being of porbeagle sharks and their ecosystem.

Author Contributions: Conceptualization, A.J.; methodology, A.J., A.O. and L.Z.; software, A.O.; validation, A.J. and L.Z.; formal analysis, A.J., A.O. and P.A.; investigation, A.J., A.O., P.A. and L.Z.; resources, L.Z.; data curation, A.J., A.O. and L.Z.; writing—original draft preparation, A.J.; writing—review and editing, A.J., A.O. and L.Z.; visualization, A.O. and L.Z.; supervision, A.J.; project administration, A.J.; funding acquisition, A.J. All authors have read and agreed to the published version of the manuscript.

Funding: This research was funded by Des Requins et Des Hommes (DRDH), Technopôle Brest-Iroise, 29280 Plouzané, France and the French Ministry for Ecological Transition and Territorial Cohesion, Arche de la Defense, 92055 Paris la Defense, France.

Institutional Review Board Statement: The study was conducted in accordance with. The French Charter on the Ethic of Animal Experimentation from both the Ministries of Research and Agriculture (Paris, France). A specific research permit for biopsy was attributed by the Direction Inter-Régionale de la Mer Nord-Atlantique, Manche-Ouest (DIRM NAMO, Rennes, France), n°: 708/2023 delivered the 6 June 2023.

Data Availability Statement: Updated of this project will be available on the https://www.researchgate.net/profile/Armelle-Jung.

Acknowledgments: We are grateful to the professional photographers who kindly accepted to share their videos: Didier Brémont, Yannick Cherel, Jérome Delafosse, Stéphane Granzotto, Yves Lefèvre, Andy Murch, Armel Ruy, the Ocean Collective team, and all the marine stakeholders websites contributors. We address a sincere thanks to Pascal Provost, director of the Nature Reserve of the Seven Islands and his whole team; as well as to the Conservatoire du Littoral, Lannion Community, and Perros Guirec harbor managers. All this work would not have been possible without the help of DRDH members (divers, drivers, and technical assistance), and the active contribution of July Plongée diving club and Didier Brémont to the ethical code of conduct and their facilitation of the field research. The authors are grateful for the comments provided by reviewers as well as colleagues' proofreading and advice.

Conflicts of Interest: The authors declare no conflicts of interest.

Appendix A

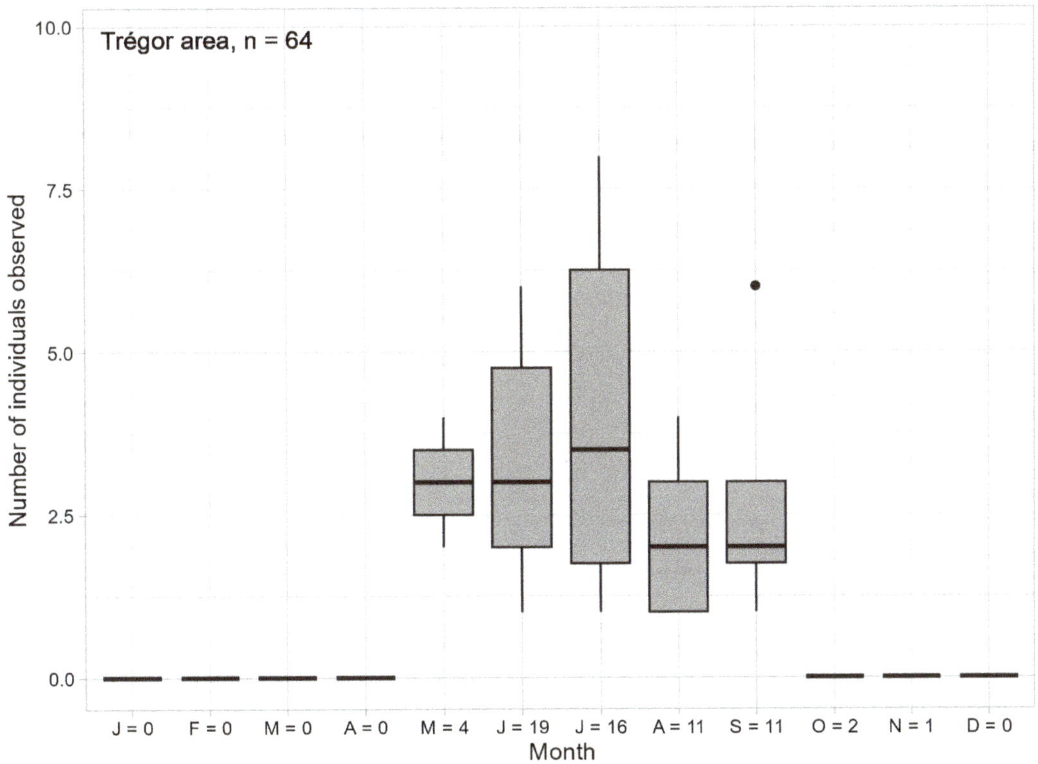

Figure A1. Distribution of the number of porbeagle shark sightings by dive per month between 2021 and 2023 (observed by DRDH), with the accumulated number of dives per month on the *x*-axis.

Figure A2. Example of maturity stage identification for the porbeagle shark 20LAMNA17 "Fanny" observed by DRDH in June 2023 (gravid) and August 2023 (postpartum).

References

1. Francis, M.P.; Natanson, L.J.; Campana, S.E. The biology and ecology of the porbeagle shark, *Lamna nasus*. In *Sharks of the Open Ocean: Biology, Fisheries and Conservation*; Blackwell Publishing: Oxford, UK, 2008; Chapter 9; pp. 105–113. [CrossRef]
2. Pade, N.G.; Queiroz, N.; Humphries, N.E.; Witt, M.J.; Jones, C.S.; Noble, L.R.; Sims, D.W. First results from satellite-linked archival tagging of porbeagle shark, *Lamna nasus*: Area fidelity, wider-scale movements and plasticity in diel depth changes. *J. Exp. Mar. Biol. Ecol.* **2009**, *370*, 64–74. [CrossRef]
3. Biais, G.; Coupeau, Y.; Séret, B.; Calmettes, B.; Lopez, R.; Hetherington, S.; Righton, D. Return migration patterns of porbeagle shark (*Lamna nasus*) in the Northeast Atlantic: Implications for stock range and structure. *ICES J. Mar. Sci.* **2017**, *74*, 1268–1276. [CrossRef]
4. Stevens, J.D. Stomach contents of the blue shark (*Prionace glauca* L.) off south-west England. *J. Mar. Biol. Assoc. U. K.* **1973**, *53*, 357–361. [CrossRef]
5. Ellis, J.R.; Shackley, S.E. Notes on porbeagle sharks, *Lamna nasus*, from the Bristol Channel. *J. Fish Biol.* **1995**, *46*, 368–370. [CrossRef]
6. Belleggia, M.; Colonello, J.; Cortés, F.; Figueroa, D.E. Eating catch of the day: The diet of porbeagle shark *Lamna nasus* (Bonnaterre 1788) based on stomach content analysis, and the interaction with trawl fisheries in the south-western Atlantic (52° S–56° S). *J. Fish Biol.* **2021**, *99*, 1591–1601. [CrossRef]
7. Campana, S.E.; Joyce, W.; Marks, L.; Hurley, P.; Natanson, L.J.; Kohler, N.E.; Jensen, C.F.; Mello, J.J.; Pratt, H.L.J.; Myklevoll, S.; et al. The rise and fall (again) of the porbeagle shark population in the northwest atlantic. In *Sharks of the Open Ocean: Biology, Fisheries & Conservation*; Camhi, M.D., Pikitch, E.K., Babcock, E.A., Eds.; Blackwell Publishing: Oxford, UK, 2008; pp. 445–461.
8. *Porbeagle (Lamna nasus) in the Northeast Atlantic*; Report of the ICES Advisory Committee; ICES: Copenhagen, Denmark, 2010; Volume 9, pp. 85–93.
9. Stevens, J.D.; Fowler, S.L.; Soldo, A.; McCord, M.; Baum, J.; Acuña, E.; Domingo, A.; Francis, M. Lamna nasus. *The IUCN Red List of Threatened Species*. 2006. Available online: https://www.iucnredlist.org/species/11200/3261697 (accessed on 10 October 2023).
10. Ellis, J.; Farrell, E.; Jung, A.; McCully, S.; Sims, D.; Soldo, A. *Lamna nasus* (Europe assessment). The IUCN Red List of Threatened Species 2015: E.T11200A48916453. Available online: https://www.iucnredlist.org/species/11200/48916453 (accessed on 11 December 2023).
11. Council Regulation (EU). Council Regulation (EU) No 23/2010 of 14 January 2010 fixing for 2010 the fishing opportunities for certain fish stocks and groups of fish stocks, applicable in EU waters and, for EU vessels, in waters where catch limitations are required and amending Regulations (EC) No 1359/2008, (EC) No 754/2009, (EC) No 1226/2009, and (EC) No 1287/2009. *Off. J. Eur. Union* **2010**, *L21*, 1–120.
12. Estes, R.D. *A Guide to Watching African Mammals: Including Hoofed Mammals, Carnivores, and Primates*; Chelsea Green Publishing: White River Junction, VE, USA, 1993.
13. Karanth, K.U.; Nichols, J.D. Estimation of tiger densities in India using photographic captures and recaptures. *Ecology* **1998**, *79*, 2852–2862. [CrossRef]
14. Aschettino, J.M.; Baird, R.W.; McSweeney, D.J.; Webster, D.L.; Schorr, G.S.; Huggins, J.L.; Martien, K.K.; Mahaffy, S.D.; West, K.L. Population structure of melon-headed whales (*Peponocephala electra*) in the Hawaiian Archipelago: Evidence of multiple populations based on photo-identification. *Mar. Mammal Sci.* **2012**, *28*, 666–689. [CrossRef]
15. Wursig, B.; Jefferson, T.A. Methods of photo-identification for small cetaceans. In *Individual Recognition of Cetaceans: Use of Photo-Identification and Other Techniques to Estimate Population Parameters*; Hammond, P.S., Mizroch, S.A., Donovan, G.P., Eds.; Reports of the International Whaling Commission, Special Issue 12; International Whaling Commission: Cambridge, UK, 1990; pp. 43–52.
16. Vincent, C.; Meynier, L.; Ridoux, V. Photo-identification in grey seals: Legibility and stability of natural markings. *Mammalia* **2001**, *65*, 363–372. [CrossRef]
17. Mazzoil, M.; McCulloch, S.D.; Defran, R.H.; Murdoch, M.E. Use of digital photography and analysis of dorsal fins for photo-identification of bottlenose dolphins. *Aquat. Mamm.* **2004**, *30*, 209–219. [CrossRef]
18. Taylor, G. *Whale Sharks, the Giants of Ningaloo Reef*; Angus & Robertson: Sydney, NSW, Australia, 1994; ISBN 0207184984/13.
19. Corcoran, M.J.; Gruber, S.H. The use of photo-identification to study the social organization of the spotted eagle ray (*Aetobatus narinari*). *Bahamas J. Sci.* **1999**, *11*, 21–27.
20. Gore, M.A.; Frey, P.H.; Ormond, R.F.; Allan, H.; Gilkes, G. Use of Photo-Identification and Mark-Recapture Methodology to Assess Basking Shark (*Cetorhinus maximus*) Populations. *PLoS ONE* **2016**, *11*, e0150160. [CrossRef]
21. Graham, R.; Roberts, C.M. Assessing the size, growth, and structure of a seasonal population of whale sharks (*Rhincodon typus*, Smith 1828) using conventional tagging and photo identification. *Fish. Res.* **2007**, *84*, 71–80. [CrossRef]
22. Couturier, L.I.E.; Jaine, F.R.A.; Townsend, K.A.; Weeks, S.J.; Richardson, A.J.; Bennett, M.B. Distribution, site affinity and regional movements of the manta ray, *Manta alfredi* (Kreft, 1868), along the east coast of Australia. *Mar. Freshw. Res.* **2011**, *62*, 628–637. [CrossRef]
23. Marshall, A.D.; Pierce, S.J. The use and abuse of photographic identification in sharks and rays. *J. Fish Biol.* **2012**, *80*, 1361–1379. [CrossRef]
24. Dudgeon, C.L.; Noad, M.J.; Lanyon, J.M. Abundance and demography of a seasonal aggregation of zebra sharks *Stegostoma fasciatum*. *Mar. Ecol. Prog. Ser.* **2008**, *368*, 269–281. [CrossRef]

25. Hewitt, A.M.; Kock, A.A.; Booth, A.J.; Griffiths, C.L. Trends in sightings and population structure of white sharks, *Carcharodon carcharias*, at Seal Island, False Bay, South Africa, and the emigration of subadult female sharks approaching maturity. *Environ. Biol. Fishes* **2018**, *101*, 39–54. [CrossRef]
26. Bettcher, V.B.; Santos, L.N.; Bertoncini, Á.A.; Silva, M.B.; Castro, A.L. Evidence of the Atlantic nurse shark (*Ginglymostoma cirratum*) population shrink at Rocas Atoll, Southwestern Atlantic. *Aquat. Conserv. Mar. Freshw. Ecosyst.* **2023**, *33*, 845–852. [CrossRef]
27. Castro, A.L.; Rosa, R.S. Use of natural marks on population estimates of the nurse shark, *Ginglymostoma cirratum*, at Atol das Rocas Biological Reserve, Brazil. *Environ. Biol. Fishes* **2005**, *72*, 213–221. [CrossRef]
28. Nakachi, K.I. Heeding the History of Kahu Manō: Developing and Validating a Pono Photo-Identification Methodology for Tiger Sharks (*Galeocerdo cuvier*) in Hawaii. Ph.D. Thesis, University of Hawaii at Hilo, Hilo, HI, USA, 2021.
29. Madon, B.; Gimenez, O.; McArdle, B.; Baker, C.S.; Garrigue, C. A New Method For Estimating Animal Abundance with Two Sources of Data in Capture-Recapture Studies. *Methods Ecol. Evol.* **2011**, *2*, 390–400. [CrossRef]
30. Holmberg, J.; Norman, B.; Arzoumanian, Z. Estimating population size, structure, and residency time for whale sharks *Rhincodon typus* through collaborative photo-identification. *Endanger. Species Res.* **2009**, *7*, 39–53. [CrossRef]
31. Lewis, R.; Dawson, S.; Rayment, W. Estimating population parameters of broadnose sevengill sharks (*Notorynchus cepedianus*) using photo identification capture-recapture. *J. Fish Biol.* **2020**, *97*, 987–995. [CrossRef]
32. *Porbeagle (Lamna nasus) in Subareas 1–10, 12, and 14 (the Northeast Atlantic and Adjacent Waters)*; Report of the ICES Advisory Committee; ICES: Copenhagen, Denmark, 2022. [CrossRef]
33. Le Person, A. *La Pêche au Gros en Baie de Lannion: Thons Rouges et Requins*; Embanner, Y., Ed.; Embanner: Fouesnant, France, 2013; 336p, ISBN 978-2916579573.
34. Siorat, F. L'évolution de la colonie de Fous de Bassan, Morus bassanus, de l'île Rouzic de 1939 à 1996. *Le Cormoran* **1998**, *47*, 173–175.
35. Chauvaud, S.; Jean, F. *Inventaire Patrimonial de la Macrofaune de L'Estran de la Réserve Naturelle des Sept-îles*; Rapport C.R.O.E.M.I, CROEMI-UBO, Université de Bretagne Occidentale: Brest, France, 1999; 22p.
36. Siorat, F.; Bentz, G. Réserve Naturelle des Sept-Îles. Activity Report. 2004. Available online: https://www.maia-network.org (accessed on 20 October 2023).
37. Martelet, G.; Gloaguen, E.; Døssing, A.; Lima Simoes da Silva, E.; Linde, J.; Rasmussen, T.M. Airborne/UAV Multisensor Surveys Enhance the Geological Mapping and 3D Model of a Pseudo-Skarn Deposit in Ploumanac'h, French Brittany. *Minerals* **2021**, *11*, 1259. [CrossRef]
38. Kalmijn, A.J. Electric and magnetic sensory world of sharks, skates, and rays. In *Sensory Biology of Sharks, Skates, and Rays*; Hodgson, E.S., Mathewson, R., Eds.; Office of Naval Research, Department of the Navy: Arlington, VA, USA, 1978; pp. 507–528.
39. Tricas, T.C. The neuroecology of the elasmobranch electrosensory world: Why peripheral morphology shapes behavior. In *The Behavior and Sensory Biology of Elasmobranch Fishes: An Anthology in Memory of Donald Richard Nelson*; Tricas, T.C., Gruber, S.H., Eds.; Springer: Dordrecht, The Netherlands, 2001; Volume 20. [CrossRef]
40. Hutchison, Z.; Sigray, P.; He, H.; Gill, A.B.; King, J.; Gibson, C. *Electromagnetic Field (EMF) Impacts on Elasmobranch (Shark, Rays, and Skates) and American Lobster Movement and Migration from Direct Current Cables*; US Department of the Interior, Bureau of Ocean Energy Management: Sterling, VA, USA, 2018. [CrossRef]
41. Serre, S.; Jung, A.; Cherel, Y.; Gamblin, C.; Hennache, C.; Le Loc'h, F.; Lorrain, A.; Priac, A.; Schaal, G.; Stephan, E. Stable isotopes reveal intrapopulation heterogeneity of porbeagle shark (*Lamna nasus*). *Reg. Stud. Mar. Sci.* **2024**, *69*, 103340. [CrossRef]
42. Domeier, M.L.; Nasby-Lucas, N. Annual re-sightings of photographically identified white sharks (*Carcharodon carcharias*) at an eastern Pacific aggregation site (Guadalupe Island, Mexico). *Mar. Biol.* **2007**, *150*, 977–984. [CrossRef]
43. Anderson, S.D.; Chapple, T.K.; Jorgensen, S.J.; Klimley, A.P.; Block, B.A. Long-term individual identification and site fidelity of white sharks, *Carcharodon carcharias*, off California using dorsal fins. *Mar. Biol.* **2011**, *158*, 1233–1237. [CrossRef]
44. Jensen, C.F.; Natanson, L.J.; Pratt, H.L., Jr.; Kohler, N.E.; Campana, S.E. The reproductive biology of the porbeagle shark (*Lamna nasus*) in the western North Atlantic Ocean. *Fish. Bull.* **2002**, *100*, 727–738.
45. Dureuil, M.; Towner, A.V.; Ciolfi, L.G.; Beck, L.A. A computer-aided framework for subsurface identification of white shark pigment patterns. *Afr. J. Mar. Sci.* **2015**, *37*, 363–371. [CrossRef]
46. Jung, A. *A Preliminary Assessment of the French Fishery Targeted Porbeagle Shark (Lamna nasus) in the Northeast Atlantic Ocean: Biology and Catch Statistics*; Standing Committee on Research and Statistics, ICCAT: Madrid, Spain, 2008; p. 11.
47. Saunders, R.A.; Royer, F.; Clarke, M.W. Winter migration and diving behaviour of porbeagle shark, *Lamna nasus*, in the Northeast Atlantic. *ICES J. Mar. Sci.* **2011**, *68*, 166–174. [CrossRef]
48. Francis, M.P.; Stevens, J.D. Reproduction, embryonic development, and growth of the porbeagle shark, *Lamna nasus*, in the southwest Pacific Ocean. *Fish. Bull.* **2000**, *98*, 41–63.
49. Penfold, L.M.; Wyffels, J.T. Reproductive science in sharks and rays. In *Reproductive Sciences in Animal Conservation, Advances in Experimental Medicine and Biology*; Comizzoli, P., Brown, J., Holt, W., Eds.; Springer: Cham, Switzerland, 2019; Volume 1200. [CrossRef]
50. Anderson, B.; Belcher, C.; Slack, J.A.; Gelsleichter, J. Evaluation of the use of portable ultrasonography to determine pregnancy status and fecundity in bonnethead shark *Sphyrna tiburo*. *J. Fish Biol.* **2018**, *93*, 1163–1170. [CrossRef]
51. Barnett, A.; Abrantes, K.G.; Stevens, J.D.; Semmens, J.M. Site fidelity and sex-specific migration in a mobile apex predator: Implications for conservation and ecosystem dynamics. *Anim. Behav.* **2011**, *81*, 1039–1048. [CrossRef]
52. Domeier, M.L. *Global Perspectives on the Biology and Life History of the White Shark*; CRC Press: Boca Raton, FL, USA, 2012.

53. Pillans, R.D.; Rochester, W.; Babcock, R.C.; Thomson, D.P.; Haywood, M.D.E. Long-Term Acoustic Monitoring Reveals Site Fidelity, Reproductive Migrations, and Sex Specific Differences in Habitat Use and Migratory Timing in a Large Coastal Shark (*Negaprion acutidens*). *Front. Mar. Sci.* **2021**, *8*, 616633. [CrossRef]
54. Hulbert, L.B.; Aires-da-Silva, A.M.; Gallucci, V.F.; Rice, J.S. Seasonal foraging movements and migratory patterns of female *Lamna ditropis* tagged in Prince William Sound, Alaska. *J. Fish Biol.* **2005**, *67*, 490–509. [CrossRef]
55. Druon, J.-N.; Campana, S.; Vandeperre, F.; Hazin, F.H.V.; Bowlby, H.; Coelho, R.; Queiroz, N.; Serena, F.; Abascal, F.; Damalas, D.; et al. Global-Scale Environmental Niche and Habitat of Blue Shark (*Prionace glauca*) by Size and Sex: A Pivotal Step to Improving Stock Management. *Front. Mar. Sci.* **2022**, *9*, 828412. [CrossRef]
56. Hennache, C.; Jung, A. Etude de la Pêcherie Palangrière de Requin Taupe de L'île D'Yeu. Brest, France: Association Pour L'étude et la Conservation des Sélaciens (APECS). 2010. Available online: https://www.researchgate.net/profile/Armelle-Jung/publication/313667107_Etude_de_la_pecherie_palangriere_de_requin_taupe_Lamna_nasus_de_l%E2%80%99Ile_d%E2%80%99Yeu_EPPARTIY/links/58a2315daca272046aafe5f1/Etude-de-la-pecherie-palangriere-de-requin-taupe-Lamna-nasus-de-lIle-dYeu-EPPARTIY.pdf (accessed on 20 April 2021).
57. Bradford, R.; Patterson, T.A.; Rogers, P.J.; McAuley, R.; Mountford, S.; Huveneers, C.; Robbins, R.; Fox, A.; Bruce, B.D. Evidence of diverse movement strategies and habitat use by white sharks, *Carcharodon carcharias*, off southern Australia. *Mar. Biol.* **2020**, *167*, 96. [CrossRef]
58. Barker, S.M.; Williamson, J.E. Collaborative photo-identification and monitoring of grey nurse sharks (*Carcharias taurus*) at key aggregation sites along the eastern coast of Australia. *Mar. Freshw. Res.* **2010**, *61*, 971–979. [CrossRef]
59. Van Tienhoven, A.M.; Den Hartog, J.E.; Reijns, R.A.; Peddemors, V.M. A computer-aided program for pattern-matching of natural marks on the spotted raggedtooth shark *Carcharias taurus*. *J. Appl. Ecol.* **2007**, *44*, 273–280. [CrossRef]
60. Sarano, F.; Girardet, J.; Sarano, V.; Vitry, H.; Preud'Homme, A.; Heuzey, R.; Garcia-Cegarra, A.M.; Madon, B.; Delfour, F.; Glotin, H.; et al. Kin relationships in cultural species of the marine realm: Case study of a matrilineal social group of sperm whales off Mauritius island, Indian Ocean. *R. Soc. Open Sci.* **2021**, *8*, 201794. [CrossRef]
61. González, M.T.; Sepúlveda, F.A.; Zárate, P.M.; Baeza, J.A. Regional population genetics and global phylogeography of the endangered highly migratory shark Lamna nasus: Implications for fishery management and conservation. *Aquat. Conserv. Mar. Freshw. Ecosyst.* **2021**, *31*, 620–634. [CrossRef]
62. Madon, B.; Le Guyader, D.; Jung, J.-L.; De Montgolfier, B.; Lopez, P.J.; Foulquier, E.; Bouveret, L.; Le Berre, I. Pairing AIS data and underwater topography to assess maritime traffic pressures on cetaceans: Case study in the Guadeloupean waters of the Agoa sanctuary. *Mar. Policy* **2022**, *143*, 105160. [CrossRef]
63. Séguigne, C.; Mourier, J.; Clua, É.; Buray, N.; Planes, S. Citizen science provides valuable data to evaluate elasmobranch diversity and trends throughout the French Polynesia's shark sanctuary. *PLoS ONE* **2023**, *18*, e0282837. [CrossRef] [PubMed]
64. Coché, L.; Arnaud, E.; Bouveret, L.; David, R.; Foulquier, E.; Gandilhon, N.; Jeannesson, E.; Le Bras, Y.; Lerigoleur, E.; Lopez, P.J.; et al. Kakila database: Towards a FAIR community approved database of cetacean presence in the waters of the Guadeloupe Archipelago, based on citizen science. *Biodivers. Data J.* **2021**, *9*, e69022. [CrossRef] [PubMed]

diversity

MDPI

Article

A Proposed Method for Assessing the Spatio-Temporal Distribution of *Carcharhinus melanopterus* (Quoy and Gaimard, 1824) in Shallow Waters Using a UAV: A Study Conducted in Koh Tao, Thailand

Andrea Di Tommaso, Sureerat Sailar, Francesco Luigi Leonetti , Emilio Sperone * and Gianni Giglio *

Laboratory of Marine Zoology and Herpetology, Department of Biology, Ecology and Earth Sciences, University of Calabria, 87036 Rende, Italy; francescoluigi.leonetti@unical.it (F.L.L.)
* Correspondence: emilio.sperone@unical.it (E.S.); gianni.giglio@unical.it (G.G.); Tel.: +39-0984492972 (E.S.)

Abstract: In this study, we propose a method for assessing the temporal and spatial distribution of *Carcharhinus melanopterus* in shallow waters using unmanned aerial vehicles (UAVs). Aerial surveys were conducted in Tien Og Bay (Koh Tao, Thailand) thrice daily (morning, afternoon, evening) along a 360 m transect at a 30 m altitude. Environmental factors, including cloudiness, sea conditions, wind, tide, and anthropogenic disturbance, were recorded for each time slot. We developed a Python/AppleScript application to facilitate individual counting, correlating sightings with GPS data and measuring pixel-based length. Abundance varied significantly across time slots ($p < 0.001$), with a strong morning preference, and was influenced by tide ($p = 0.040$), favoring low tide. Additionally, abundance related to anthropogenic disturbance ($p = 0.048$), being higher when anthropogenic activity was absent. Spatial distribution analysis indicated time-related, sector-based abundance differences ($p < 0.001$). Pixel-based length was converted to Total Length, identifying juveniles. They exhibited a strong sector preference ($p < 0.001$) irrespective of the time of day. Juvenile abundance remained relatively stable throughout the day, constituting 94.1% of afternoon observations. Between 2020 and 2022, an underwater video survey was conducted to determine the sex ratio of the individuals. Only females and juveniles were sighted in the bay.

Keywords: unmanned aerial vehicles; sharks; ecology

1. Introduction

Carcharhinus melanopterus (Quoy and Gaimard, 1824) is one of the most abundant sharks inhabiting islands, atolls, and coastal waters from the Red Sea to the Indian Ocean and the central Pacific [1–3].

The high population density observed when direct threats are absent suggests that it exerts a certain influence on coral ecosystems [2,4]. In 2020, it was classified as vulnerable by IUCN mainly because of the shrinkage of the population across the last three generations due to fishing and habitat destruction.

Carcharhinus melanopterus is a medium-sized shark (<1.80 m TL) that reaches maturity between 0.9 and 1.34 m TL [2,5,6]. The reproductive cycle is biennial [2] and is characterized by placental viviparity. Two–four pups measuring 0.3–0.5 m TL [5,7] are delivered after 7 to 16 months' gestation [7,8]. In the Great Barrier Reef, it has been estimated that maturity is reached around 4.2 years for males and 8.5 years for females. Captive specimens can reach a life span of 25 years [6]. The duration of a generation is estimated to be 14.5 years.

Carcharhinus melanopterus inhabits shallow waters even less than one meter in depth, but has been found in water up to a depth of 75 m [5]. In addition to coral reefs, it also colonizes turbid waters and mangroves [7,9]. The coastal habits make it an easy species to observe and consequently a possible tourist attraction, which is also due to its low aggressiveness towards humans, making up around 3% of the total recorded shark attacks [10]. It feeds mainly on small fish, crustaceans, and mollusks, but also snakes [11], birds, and even rats in specific locations such as Australia, Seychelles, and French Polynesia [10,12].

The species is often observed forming aggregations [13] and has a high site-fidelity [2,12]. This has led some authors to suppose that the patterns with which aggregations occur may be stable over time [13]. The site-fidelity makes *Carcharhinus melanopterus* an ideal model with which to study social structure and aggregation patterns in elasmobranchs, and to determine by which factors they are influenced. Currently, there is no evidence of territoriality in elasmobranchs [14]; however, hierarchies based on size may exist. Some scholars have hypothesized that fin markings of some species may play a role in the species-specific recognition and size assessment of other individuals in sharks such as *Triaenodon obesus* (whitetip reef shark) and *Carcharhinus melanopterus* [15].

This study aims to contribute to the knowledge of the ecology of the Blacktip reef shark, particularly in Tien Og Bay (Thailand). In order to optimize data collection in the natural environment, a method for assessing the temporal and spatial distribution of *Carcharhinus melanopterus* in shallow waters using unmanned aerial vehicles (UAVs) is proposed.

Over the past decade, drones, BRUVS, and other new technologies have become a popular tool for shark research [16,17]. In particular, the rapid proliferation of the drone technology has enabled new studies in the field of shark research, focused mainly on shark hazard reduction [18], shark predation events [19], shark behavior, and social interactions [20,21], but also pelagic shark aggregations [22]. Drone technology has also been successfully used to study the ecology and behavior of reef sharks [23], where field studies often suffer from objective difficulties in collecting data. In fact, shallow reefs are generally only accessible by foot at low tides or by boat at high tides; this makes using common abundance or behavior survey techniques difficult to use effectively.

In these shallow reef environments, drones currently provide one of the only means to study shark behavior, distribution, and abundance. The use of drones in these environments, however, is relatively novel, and there are few studies that have relied on this approach to obtain relevant data. Furthermore, it is also necessary to work on data collection and analysis methodologies to develop protocols that allow for optimizing costs and times and obtaining quality data [24].

Study Area

The research was conducted in Tien Og Bay (Figure 1), located in the southeast of Koh Tao, Thailand. Koh Tao is a 21 km^2 granitic island situated 72 km from the mainland. The waters surrounding the island are home to coral communities that, under the most favorable conditions, can form a fringing reef [25]. Tien Og Bay is commonly referred to as Shark Bay due to the relative ease with which *Carcharhinus melanopterus* can be observed.

Figure 1. Tien Og Bay, located in the southeast of Koh Tao, Thailand.

The bay has an extensive beach of fine coral sand in a North–North-West position, in front of the Haad Tien resort. On both rocky sides, other resorts are found, from which it is possible to access the bay. The seabed, starting from the Haad Tien Resort beach, is initially sandy. After a few meters, it presents the first isolated coral structures which gradually become more aggregated. Living corals alternate with dead coral rubble, which sometimes serve as a substrate for macroalgae. Granite boulders are sparsely distributed and colonized by corals of the genus *Porites* spp., forming structures known as "mini-atolls" due to the characteristic morphology they assume when the low depth of the water limits their growth in height. Proceeding out to sea—in the South–South-East direction—after about 200 m from the beach, the seabed reaches a depth of 2–3 m and a thick layer of dead Acropora rubble can be found, alternately colonized by the macroalgae *Turbinaria* spp., living Acropora, and sponges. The main tourist activity is SCUBA diving. The island has 67 diving schools [26], making it the second largest location in the world for the number of SCUBA certifications issued per year [27]. Diving, snorkeling, and marine traffic exert a significant pressure on the system, sometimes physically damaging shallow-water coral structures [28,29]. In recent years, some Thai beaches have been closed to the public to allow ecosystems to recover from the strong anthropogenic pressure. The case of Maya Bay in Phi Phi Islands is remarkable, which, a few months after its closure to the public in 2018, registered a massive aggregation of *Carcharhinus melanopterus*.

The following shark species have historically been present in Koh Tao coastal waters:

- *Rhincodon typus* (whale shark);
- *Carcharhinus melanopterus* (Blacktip reef shark);
- *Carcharhinus amblyrhynchos* (Gray reef shark);
- *Carcharhinus leucas* (Bull shark);
- *Triakis semifasciata* (Leopard shark).

In recent years, only *Carcharhinus melanopterus* and, exceptionally, the whale shark *Rhincodon typus* have been found [30].

2. Materials and Methods

The study was carried out between 19 August 2021 and 9 September 2021. A DJI Mini 2 drone, DJI Official Store Bangkok (Bangkok, Thailand) was used in manual flight for data collection. The DJI Mini 2 is equipped with a 1/2.3″ 12MP CMOS sensor and a 4.49 mm lens (equivalent to 24 mm in full-frame format), with an 83° viewing angle and a maximum aperture of f/2.8.

For data collection, a transect of about 360 m was covered at an average speed of 2 m/s 4 times (2 round trips) for each time slot 3 times a day (07:00, 13:00, 18:00), except in cases of extremely bad weather conditions. Videos were recorded with a resolution of 2.7 K (2720 × 1530) at 30 fps. In total, we recorded (details in Table 1) the following:

- 07:00—20 videos, from 20 August to 9 September 2021;
- 13:00—16 videos, from 19 August to 4 September 2021;
- 18:00—14 videos, from 19 August to 4 September 2021.

For each time slot, the following environmental factors were recorded. A discrete numerical value was assigned to each condition as follows:

1. Cloudiness—0: clear (0–2 Oktas); 1: partly cloudy (3–5 Oktas); 2: cloudy (6–8 Oktas); 3: rain (after/before).
2. Wind—0: absent (0 m/s wind speed); 1: weak (1–2 m/s wind speed); 2: medium (3–5 m/s wind speed); 3: strong (5–10 m/s wind speed); 4: very strong (>10 m/s wind speed).
3. Tide—0: low; 1: medium; 2: high.
4. Sea condition—0 (0–1 Beaufort scale): flat; 1: quiet (2–3 Beaufort scale); 2: almost quiet (4 Beaufort scale).
5. Anthropogenic disturbance—0: absent; 1: snorkeling/kayaking; 2: motorboat.

Table 1. Activity table for each sampling day. W and R, respectively, represent NA due to wind or rain.

Date	Morning Transect Time	Afternoon Transect Time	Evening Transect Time
19 August 2021		13:10	18:08
19 August 2021	06:54	13:06	18:07
21 August 2021	07:06	13:04	18:02
22 August 2021	07:07	13:00	18:04
23 August 2021	07:06	12:55	17:59
24 August 2021	06:59	13:10	W
25 August 2021	07:15	13:04	18:07
26 August 2021	07:09	13:12	17:42
27 August 2021	07:05	12:53	17:59
28 August 2021	07:11	13:07	18:06
29 August 2021	09:23	13:09	18:03
30 August 2021	06:57	13:09	17:59
31 August 2021	07:08	R	R
01 September 2021	08:50	13:05	17:59
02 September 2021	07:03	13:23	W
03 September 2021	07:03	13:04	18:03
04 September 2021	08:18	13:10	18:04
05 September 2021	07:05		
06 September 2021	07:15		
07 September 2021	R		
08 September 2021	07:10		
09 September 2021	07:09		

Considering the presence of swimmers, boats, and infrastructures, a height of 30 m was chosen in accordance with Thai legislation about drones, which requires a minimum distance of 30 m from people, vehicles, and infrastructures.

The starting point was established in accordance with the distance of at least 30 m from the nearby resorts. The end point was chosen based on the most easily identifiable visual landmark on the other side of the bay: the white building of the Jamahkiri Resort.

Considering the focal length of the drone's lens and the dimensions of the sensor, it has been estimated that, from a height of 30 m, an image corresponding to a width of 41 m on the ground is captured.

The following equation represents the proportional relationship between the length of a line in the real world and the length of a line in pixels on an image:

$$\text{RealWorldLength} = \text{GSD} \times \text{LengthInPixels}$$

The Ground Sampling Distance (GSD), expressed in px/m, indicates how many meters a single pixel in the image corresponds to. The GSD was calculated through the following formula:

$$\text{GSD} = \frac{\text{SensorWidth} \times \text{Altitude}}{\text{FocalLength} \times \text{ImageWidth}} = \frac{6.17\text{ mm} \times 30\text{ m}}{4.49\text{ mm} \times 2720\text{ px}} = 0.015\text{ m/px}$$

In our case, we wanted to estimate the entire scene width captured by the sensor; therefore, LengthInPixel = ImageWidth = 2720 px. Given that, the following is obtained:

$$\text{WidthCameraFootprint} = \text{GSD} \times \text{LengthInPixel} = 0.015\text{ m/px} \times 2720\text{ px} = 41\text{ m}$$

2.1. Temporal Data Analysis and Environmental Factors

We flew the transect for two round trips. Each round trip included one leg in the forward direction (labeled A1 and A2, respectively) and one leg in the return direction (labeled R1 and R2), resulting in a total of four legs for the two round trips.

For each video recorded between 20 August 2021 and 04 September 2021, observed individuals were counted for each leg of the transect. The results were grouped by time slot. For each time slot, we created a table reporting the date, actual start time of the video recording, environmental factors (cloudiness, wind, tide, sea conditions, anthropogenic disturbance), and the number of sharks observed in each leg of the transect, labeled nA1, nR1, nA2, and nR2. Table 2 shows the data collected in the morning between 20 August 2021 and 04 September 2021.

Table 2. Data collected in morning time between 20 August 2021 and 04 September 2021: environmental factors and observations for each day. nA1, nR1, nA2, and nR2 represent the observations in each leg of the transect; nMax represents the highest number of observations among the legs of the transect ('abundance'); nTot represents the sum of observations across all the legs of the transect ('activity').

Date	Time	Weather (0–3)	Wind (0–4)	Tide (0–2)	Sea (0–2)	Disturbance (0–2)	nA1	nR1	nA2	nR2	nMax	nTot
20 August 2021	06:54	1	1	0	1	1	20	23	34	32	34	109
21 August 2021	07:06	2	1	0	2	0	13	11	13	14	14	51
22 August 2021	07:07	2	1	0	2	0	7	7	6	10	10	30
23 August 2021	07:06	1	2	0	1	0	22	26	32	31	32	111
24 August 2021	06:59	0	2	0	0	0	8	8	10	16	16	42
25 August 2021	07:15	3	3	1	1	0	3	4	5	4	5	16
26 August 2021	07:09	3	3	1	2	0	4	4	4	6	6	18
27 August 2021	07:05	3	1	1	1	0	6	11	13	18	18	48
28 August 2021	07:11	3	2	1	2	0	5	2	10	6	10	23
29 August 2021	09:23	2	0	0	1	0	28	24	24	16	28	92
30 August 2021	06:57	0	0	1	1	0	0	3	7	4	7	14
31 August 2021	07:08	2	0	0	1	1	4	5	5	4	5	18
01 September 2021	08:50	2	2	0	1	1	3	1	3	3	3	10
02 September 2021	07:03	2	1	0	0	1	12	11	20	8	20	51
03 September 2021	07:03	2	1	0	0	0	9	17	16	13	17	55
04 September 2021	08:18	2	0	0	0	0	NA	NA	NA	NA	NA	NA

We added a column labeled 'nMax', representing the highest number of observations among the legs of the transect, which we considered as abundance. We counted the individuals that entered the screen as soon as they were visible. The linear movement of the UAV at a speed of 2 m/s prevented double counting the same individuals within a leg. A column labeled 'nTot' was added, representing the sum of observations across all the legs of the transect, which we considered the animal's activity. Animal activity is generally defined as the amount of time an animal spends in motion [22]. Most shark species remain in motion throughout their entire lives. When measuring spatial and temporal niche partitioning among a species—such as through camera trapping—ecologists consider every sighting as 'activity' regardless of whether it involves the same or different individuals. Therefore, in our case, nTot may include double-counted individuals, as the drone moving back and forth along a line may encounter the same individuals multiple times. We used this value to compare it with the abundance.

Abundance and activity values, grouped by the 3 time slots, were analyzed to detect statistical significance using the Kruskal–Wallis test.

Abundance and activity values were then aggregated by environmental factors' conditions (cloudiness, wind, tide, sea condition, anthropogenic disturbance), and another Kruskal–Wallis test was performed to detect statistical significance among these groups.

2.2. *Spatial Data Analysis*

For the spatial distribution analysis, we needed to correlate the GPS position of the drone with the sharks observed. The DJI MINI 2 records basic telemetry in the subtitles embedded in the video files. Data such as camera settings, GPS coordinates (10 m resolution), distance from the pilot, height, and speed are recorded for each second of the video.

We used ffmpeg to extract subtitles from the videos, generating the corresponding SRT files. To aid in tracking sharks and correlating their observation times in the video with the SRT file records, we developed a hybrid Python/AppleScript application. The output of this application is a table for each video, where each row represents a shark observation and includes GPS and geometrical data.

Figure 2 illustrates the Python version 3.9.0/AppleScript version 2.8 application's interface. The application launches QuickTime Player, allowing the operator to navigate the video. When a shark is observed, the operator can use the mouse to draw a line on the shark. The application records the video time in seconds of the observation and other geometrical data. At the end of the process, a CSV file is generated for each video (Table 3), containing the following fields:

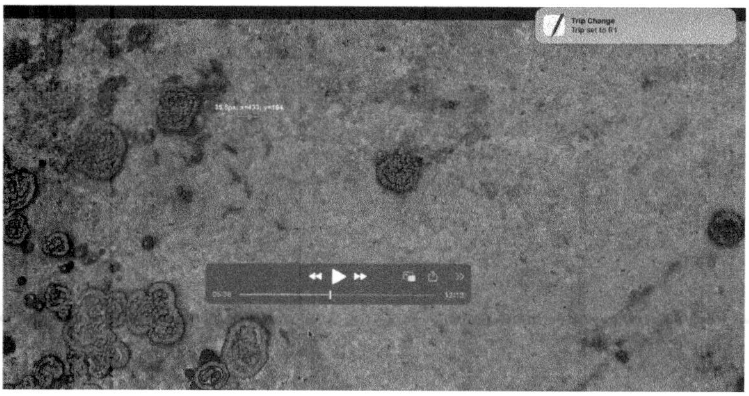

Figure 2. Python/AppleScript application's interface. The red line indicates the shark's length expressed in pixels.

- Leg of the transect (A1: first journey; R1: first return; A2: second journey; R2: second return);
- Video time in seconds of the observation;
- Position on the screen (x,y);
- Length of the line drawn on the individual in pixels;
- Sector, representing the actual transect sector, as explained below;
- GPS coordinates.

Table 3. Example of a CSV table produced by the Python/AppleScript application for a single video.

Transect Leg	Time	Screen_x	Screen_y	Length	Sector	Shark_Gps_Long	Shark_Gps_Lat
A1	120.6	653	72	33.9	6	99.8333	10.0646
A2	371.6	299	136	25.1	1	99.8312	10.0641
A2	476.4	382	28	13.9	7	99.8334	10.0646
A2	512	611	86	22.8	9	99.8342	10.0648
R2	526.2	426	678	20.7	9	99.8342	10.0648
R2	558.2	260	621	16.4	7	99.8337	10.0647

To conduct statistical tests, we divided the transect into 9 sectors, each measuring 40 m in length and approximately 41 m in width (corresponding to the camera footprint width calculated earlier). Figure 3 provides a georeferenced map created with QGIS, illustrating the layout of these sectors.

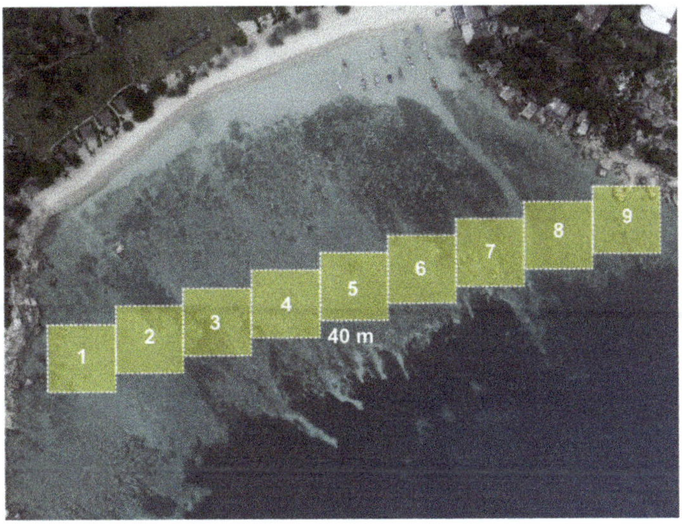

Figure 3. Map of the 9 sectors (40 m × 41 m) in which the transect was divided.

We used R to compile the tables generated by the Python/AppleScript application into three tables, each referred to as a 'sector-abundance table'. These tables document the number of observations within each sector for each respective time slot (Table 4). For the spatial distribution analysis, only one leg of the transect was used. Leg A1 was used for the morning, while the leg corresponding to nMax was used for the afternoon and evening.

Table 4. The 3 sector-abundance tables, one for each time slot, reporting the number of sharks observed and measured.

Sector Number	Sector Abundance 07:00	Sector Abundance 13:00	Sector Abundance 18:00
1	29	2	3
2	13	0	3
3	5	0	3
4	36	2	6
5	43	1	10
6	31	2	3
7	22	9	13
8	33	1	10
9	22	0	3

A Chi-squared test was conducted on each table to assess its distribution. The 3 Abundance-Sectors tables were compared using the Kruskal–Wallis test to assess the significance of the distribution among the 3 time slots.

2.3. Population Structure Analysis

The length of the lines drawn on the individuals using the developed Python/AppleScript application was used to categorize the population into two groups: juveniles and adults.

If we plot the length values of all the recorded sharks (Figure 4), we can observe that nearly 80% of the sharks' lengths fall within the range of 30 to 50 pixels. Between 30 and approximately 18 pixels, the curve steepens due to the scarcity of individuals within this range (less than 7%). Below 18 pixels, the curve gradually levels off.

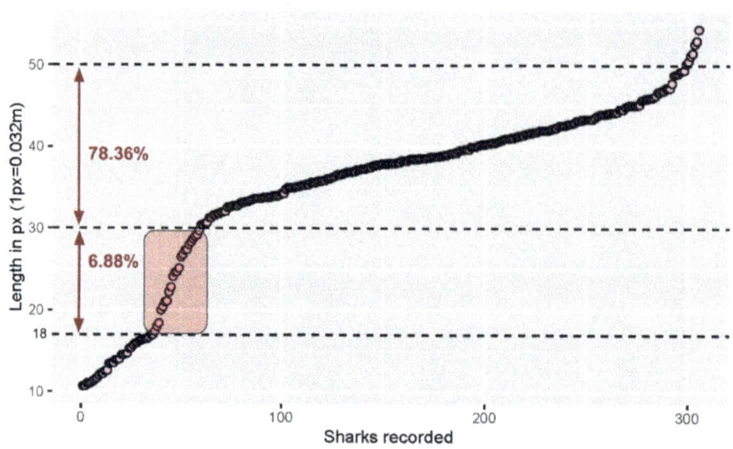

Figure 4. Length value of all the recorded sharks expressed in pixels.

Length values, measured from an aerial perspective, may be considered the PCL (Pre-Caudal Length) due to the limited visibility of the caudal fin when observed from 30 m above and under 0.5–1 m of water.

Measuring errors can potentially occur with this method due to subjectivity and image distortion. As a result, these length measurements on the screen primarily serve a comparative purpose and are influenced by the screen size on which the software is employed. For this study, we utilized a 13.3-inch screen with an effective resolution of 2560 × 1600. The actual resolution was scaled down by the Operating System to 1280 × 800. To determine the corresponding value in meters for 1 pixel, we calculated the Ground Sample Distance (GSD) based on the actual screen resolution. In our case, the obtained GSD value is as follows:

$$\text{GSD} = \frac{\text{SensorWidth} \times \text{Altitude}}{\text{FocalLength} \times \text{ImageWidth}} = \frac{6.17\ \text{mm} \times 30\ \text{m}}{4.49\ \text{mm} \times 1280\ \text{px}} = 0.032\ \text{m/px}$$

A single pixel on the screen, therefore, equals 0.032 m.

From the literature, we know that *C. melanopterus* reaches maturity starting from a Total Length of 93 cm [7].

We can calculate the corresponding PCL using a conversion formula for a morphologically similar species (*Carcharhinus brachyurus*) [31]:

$$\text{TL} = a + b * \text{PCL} = 10.270\ \text{cm} + 1.289 * \text{PCL}$$

where a and b are the conversion factors reported for *Carcharhinus brachyurus*.

Therefore, we have derived the following expression for the PCL:

$$\text{PCL} = \frac{\text{TL} - \text{a}}{\text{b}} = \frac{93\ \text{cm} - 10.270\ \text{cm}}{1.289} = 64.18\ \text{cm}$$

According to this, the minimal PCL for maturity is about 64.18 cm.

Then, we converted the obtained PCL back into pixels using the display's GSD and obtained a length of 20.06 pixels:

$$\text{LengthInPixels} = \frac{64.18\ \text{cm}}{\text{GSD}} = \frac{64.18\ \text{cm}}{3.2\ \text{cm/px}} = 20.06\ \text{px}$$

Therefore, individuals with a length less than 20 pixels were considered juveniles.

This hypothesis is corroborated by a boxplot of the measured lengths (Figure 5). The plot shows all the values below 20 pixels as outliers, where the length < Q1 − 1.5 × IQR.

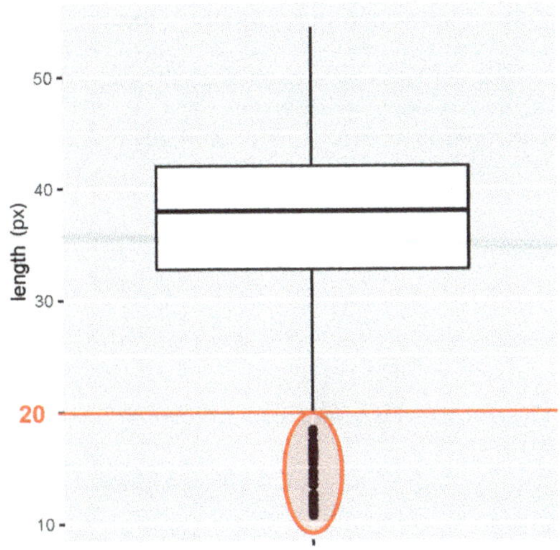

Figure 5. Boxplot of the measured lengths. Values below 20 px are considered outliers (red circle and line).

Once juveniles were separated from adults, we analyzed the abundance ratio of juveniles/adults among the 3 time slots. After that, we built 3 new sector-abundance tables only for juvenile individuals, and we performed a Kruskal–Wallis test on these 3 groups in order to identify sector preferences in juveniles.

2.4. Sex Ratio Surveys

From 2020 to 2022, we conducted an underwater video survey to determine the sex of the individuals (Table 5).

Table 5. Data from 2020–2022 underwater video surveys in order to determine the sex of the adult individuals.

Date	Time	Females	Males
07 March 2020	08:30	6	0
08 March 2020	09:00	2	0
08 March 2020	18:00	1	0
17 September 2020	09:00	1	0
05 July 2021	09:00	5	0
06 July 2021	10:00	3	0
21 September 2021	10:00	2	0
21 September 2022	09:00	1	0
23 September 2022	07:30	0	0
24 September 2022	07:00	4	0
26 September 2022	08:30	5	0
28 September 2022	08:00	5	0
Total		35	0

3. Results

3.1. Abundance and Activity in Relation to the Time Slots

The Kruskal–Wallis test performed on the abundance values (nMax) grouped by the three time slots evidenced a highly statistically significant relationship between time and abundance (kw = 26.66; $p < 0.001$) (Figure 6a). The abundance is higher at 07:00 am. Similarly, we grouped activity values (nTot) by the three time slots (07:00, 13:00, 18:00), creating three groups. We performed a Kruskal–Wallis test on these groups. The result evidenced a high statistically significant relationship between time and activity (kw = 29.04; $p < 0.001$) (Figure 6b). The activity is higher at 07:00 am.

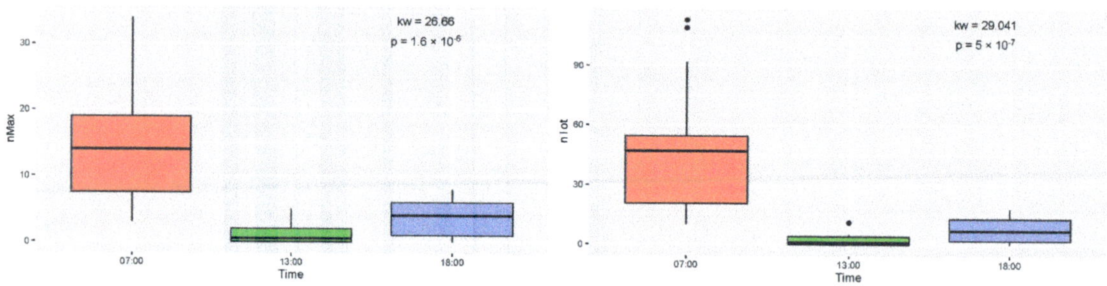

(**a**) (**b**)

Figure 6. Abundance and activity of the sharks during daytime: (**a**) time vs. abundance; (**b**) time vs. activity.

3.2. Abundance and Activity in Relation to the Environmental Factors

In this section, we present the results obtained performing the Kruskal–Wallis test on the groups formed by aggregating abundance (nMax) and activity (nTot) by the classes of values that describe each environmental factor.

3.2.1. Tide

We found a statistically significant relationship (kw = 6.42; $p = 0.04$) between the tide height and the abundance (nMax). In particular, sharks appear to be more abundant during low tide conditions (Figure 7a).

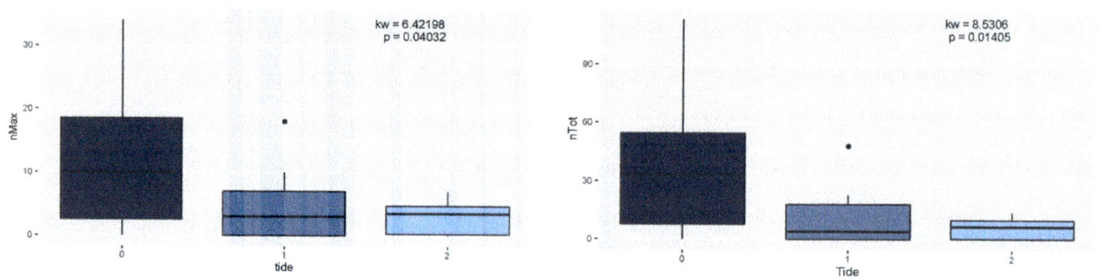

(**a**) (**b**)

Figure 7. Abundance and activity of the sharks according to tides: (**a**) tide vs. abundance; (**b**) tide vs. activity. Legend: 0 = low tide; 1 = medium tide; 2 = high tide.

Also, activity (nTot) shows a statistically significant relationship with the tide (kw = 8.53; $p = 0.014$) (Figure 7b). In particular, sharks are more active during low tide conditions.

3.2.2. Wind

A nearly significant relationship is observed (kw = 9.18; p = 0.057) between abundance (nMax) and wind: abundance is higher under lower wind conditions (0 and 1) (Figure 8a).

Similar results were found for interceptability (nTot), which is significantly associated with low wind conditions (0 and 1) (Figure 8b).

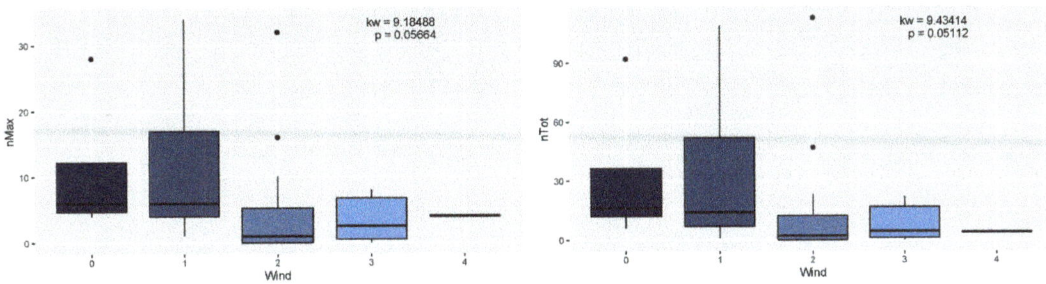

(a) (b)

Figure 8. Abundance and activity of the sharks according to wind conditions: (**a**) wind vs. abundance; (**b**) wind vs. activity. Legend: 0: absent (0 m/s wind speed); 1: weak (1–2 m/s wind speed); 2: medium (3–5 m/s wind speed); 3: strong (5–10 m/s wind speed); 4: very strong (>10 m/s wind speed).

3.2.3. Anthropogenic Disturbance

Abundance (nMax) is significantly correlated with anthropogenic disturbance (kw = 6.05; p = 0.048) (Figure 9a), as is activity (nTot) (kw = 6.8; p = 0.033) (Figure 9b). Sharks are more abundant and active under lower levels of anthropogenic disturbance.

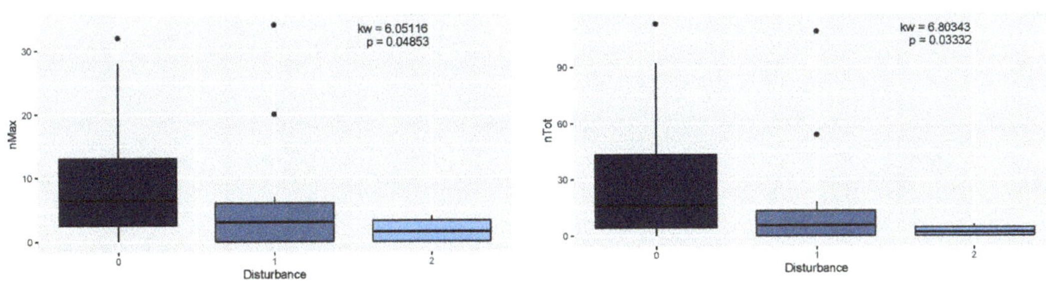

(a) (b)

Figure 9. Abundance and activity of the sharks and anthropogenic disturbance: (**a**) anthropogenic disturbance vs. abundance; (**b**) anthropogenic disturbance vs. activity. Legend: 0: absent; 1: snorkeling/kayaking; 2: motorboat.

3.2.4. Sea Condition

Throughout the observed period, wave motion within the bay remained consistently low, with occasional mild fluctuations. The graphical analysis suggests a potential correlation between the absence of substantial wave motion and both abundance (Figure 10a) and activity (Figure 10b). Nevertheless, it is important to note that these relationships did not attain statistical significance.

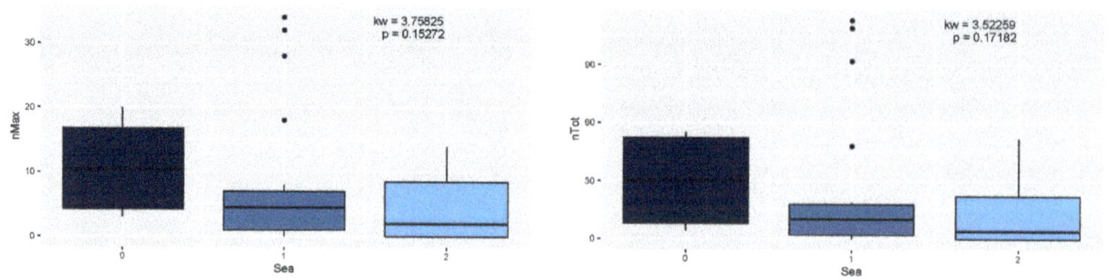

(**a**) (**b**)

Figure 10. Abundance and activity of the sharks and sea conditions: (**a**) sea condition vs. abundance; (**b**) sea condition vs. activity. Legend: 0 (0–1 Beaufort scale): flat; 1: quiet (2–3 Beaufort scale); 2: almost quiet (4 Beaufort scale).

3.2.5. Cloudiness

For both abundance (nMax) and activity (nTot), the graphs exhibit a discernible association with cloud cover, with higher values of abundance and interceptability occurring under cloudy conditions. Nevertheless, it is important to emphasize that these associations do not achieve statistical significance (abundance in Figure 11a; activity in Figure 11b).

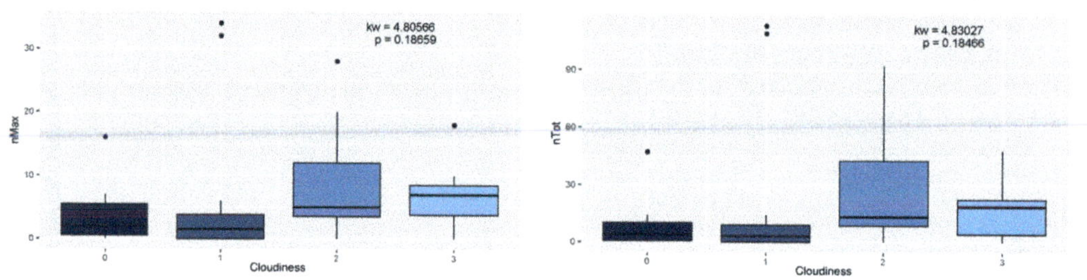

(**a**) (**b**)

Figure 11. Abundance and activity of the sharks and cloudiness: (**a**) cloudiness vs. abundance; (**b**) cloudiness vs. activity. Legend: 0: clear (0–2 Oktas); 1: partly cloudy (3–5 Oktas); 2: cloudy (6–8 Oktas); 3: rain (after/before).

3.3. Spatial Distribution

The chi-square test performed on the three sector-abundance tables (Table 4) revealed that for all three time slots, the distribution of individuals among the nine sectors significantly deviates from a normal distribution. Specifically, we obtained the following:

- Abundance-sector table at 07:00, $p < 0.001$;
- Abundance-sector table at 13:00, $p < 0.001$;
- Abundance-sector table at 18:00, $p = 0.007$.

The graphical representation of the three tables can be seen, respectively, in Figure 12a–c. The Kruskal–Wallis test across the three sector-abundance tables yielded kw = 19.91 and $p < 0.001$, indicating statistical significance in the relationship between time slot and spatial distribution. A composite graph depicting the spatial distribution across the three time slots can be seen in Figure 12d.

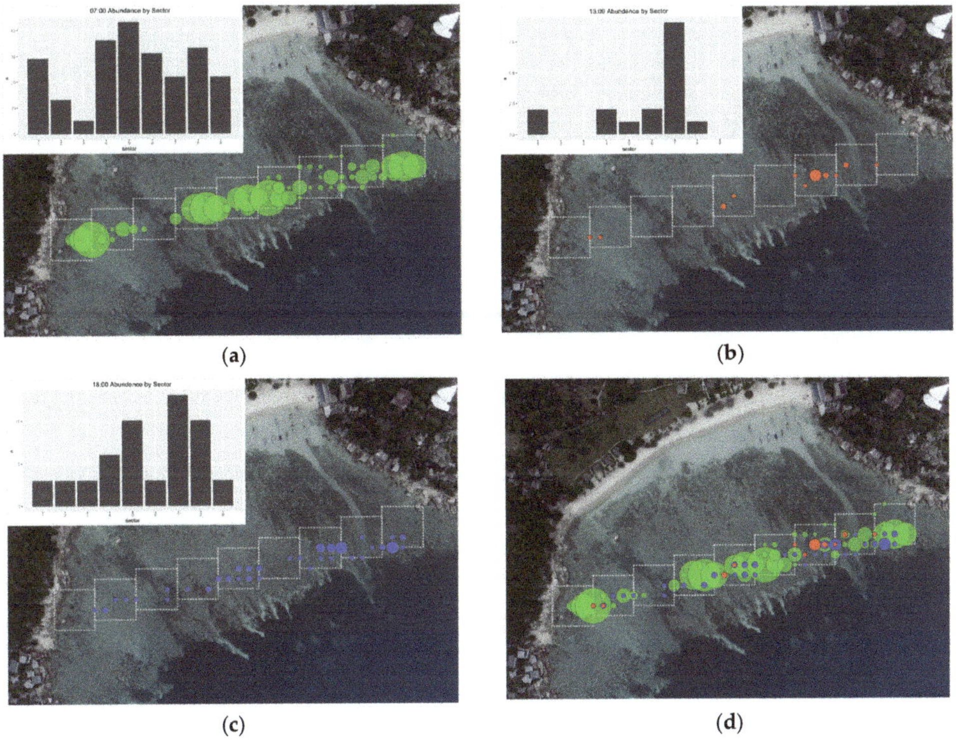

Figure 12. Abundance of the sharks during daytime in the different sectors. (**a**) Abundance by sector at 07:00; (**b**) abundance by sector at 13:00; (**c**) abundance by sector at 18:00; (**d**) composite representation of the abundance by sector: 07:00 (green), 13:00 (red), 18:00 (blue). White dotted represent the 9 sectors (40 m × 41 m) in which the transect was divided.

3.4. Population Structure and Spatio-Temporal Distribution of Juveniles

Once we have separated the juveniles from the adults, we can plot their abundance ratio across different time slots (Figure 13). In the morning, only 3.4% of the recorded individuals were juveniles. In the afternoon, almost all individuals were juveniles. In the evening, about 28% of the individuals were juveniles.

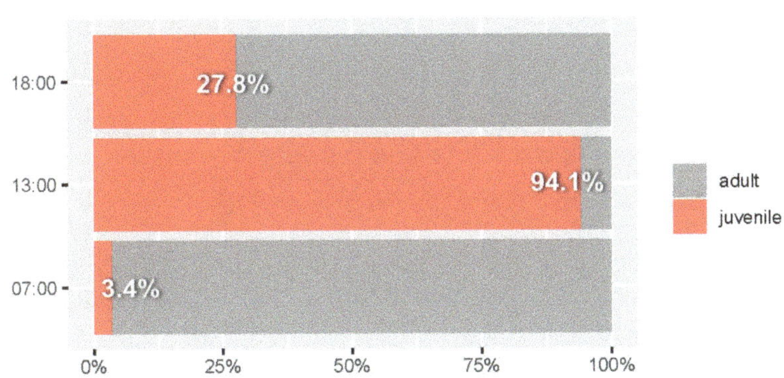

Figure 13. Juveniles/adults abundance ratio within the 3 time slots (juveniles in red).

Regarding the spatial distribution of juveniles, when conducting the Kruskal–Wallis test for the three time slots and their respective tables (abundance, juveniles, sectors), it is evident that there is no significant relationship between the occupied sectors and time slot for juveniles (kw = 1.77; $p = 0.410$). In fact, they are distributed similarly across all three time slots, with a significant preference for sector number 7 ($p < 0.001$), as can be seen in Figure 14.

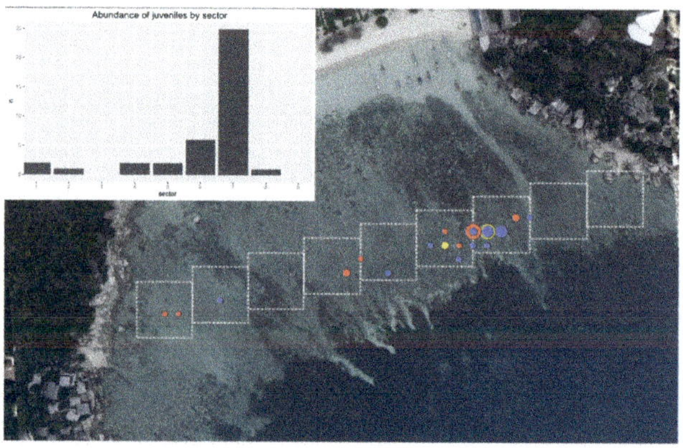

Figure 14. Composite graph depicting the distribution of juveniles across the three time slots. Legend: 07:00 (yellow), 13:00 (red), 18:00 (blue); white dotted represent the 9 sectors (40 m × 41 m) in which the transect was divided.

3.5. Sex Ratio

From 2020 to 2022, we conducted an underwater video survey to determine the sex of the individuals. Among the 35 adult individuals recorded, no male sharks were recorded; only females were present in the bay (Figure 15).

Figure 15. Some of the adult individuals observed underwater between 2020 and 2022. All observed individuals were females. (**a**) Specimen observed on 05 July 2021; (**b**) specimen observed on 17/09/2020; (**c**) specimen observed on 23 September 2022; (**d**) specimen observed on 06 July 2021; (**e**) specimen observed on 24 September 2022; (**f**) specimen observed on 17 September 2020.

4. Discussion

We proposed a method to study the spatio-temporal distribution and population structure of *Carcharhinus melanopterus* using UAVs in shallow waters. Although the study's short duration limits the conclusiveness of our results, they are still relevant to the specific time period during which the research was conducted. The proposed method demonstrates the potential of using an entry-level UAV with a minimal budget to collect and analyze data in a semi-automatic way, using the support of a desktop application to count, track, and measure the individuals for further population structure analyses.

UAV surveys provide the advantage of studying aquatic animals without causing interaction or disturbance, provided that basic requirements are met [24]. However, this approach is limited by factors such as local regulations, battery life, the effort required from UAV operators, and, eventually, extensive video analyses. In this study, a fully charged battery allowed for two round trips across a 360 m transect at a speed of 2 m/s, resulting in 12 min of video recording for each of the three time slots. Finer time scales can be achieved with multiple batteries and/or a charger, but the storage and effort for video analysis must also be considered.

Although automated flight is widely available on consumer drones or through third-party mobile apps, it is not allowed in countries such as Thailand, so it was not taken into consideration.

Recorded adult sharks could be easily spotted in the videos by human operators under all conditions encountered during the study, due to their continuous movements and thanks to shallow water. However, manual data extraction from videos becomes impractical with extensive recordings. While ML/AI techniques have proven useful for detecting sharks under specific conditions [25,26], they were challenging to implement in this study due to the varied substrate (rocks, corals, sand), superficial water movements, and variable light. Moreover, due to the presence of swimmers and boats, we had to maintain a height of 30 m according to local regulations. This resulted in a strip transect 40 m wide that, considering the length of the animals to detect, would be computationally demanding to analyze using Object Tracking. Due to the wide transect strip, juveniles were harder to spot and probably impossible to detect using ML/AI.

Considering a UAV flying a transect and the continuous shark motion, algorithms like Optical Flow could initially detect pixel groups that deviate from the UAV's linear motion, allowing the application of Object Detection only to specific image regions. Combining these techniques, Object Tracking can be theoretically achieved, preventing double counting and providing insights into the motion trajectory and speed of the animals.

The correlation between high abundance/activity and low anthropogenic disturbance suggests that human activities may exert some pressure on this species.

We found high abundance/activity correlated with low tides and time of day, with a high preference for morning time. In 9 h and 52 min of total video recordings, no bursts, sudden movements, or feeding behaviors were recorded. All individuals roamed at a constant speed all the time. This may suggest the bay is not a feeding area for adults, at least during day time. Thermoregulatory behavior in shallow bays to enhance embryo development has been extensively reported for coastal shark species [27]. In our case, this hypothesis is corroborated by the underwater survey that resulted in only females being observed.

Some sectors presented more abundance/activity than others; this may be due to local temperature, anthropic disturbance, or other parameters not recorded in this study.

Juveniles' spatio-temporal analysis revealed that they constituted a consistent percentage of the individuals at 18:00 and 13:00 due to the low presence of adults in these time slots. *Carcharhinus melanopterus* juveniles are characterized by limited motion [28]; they remain around nursery areas at all times, while adults tend to leave the bay when conditions change. Their distribution was similar across all three time slots, with a significant preference for sector number 7, for which we do not have enough data to speculate about.

Nursery areas for coastal sharks have been defined by three primary criteria for newborn or young-of-the-year individuals [29]: (1) density in the area is greater than the mean density over all areas; (2) site fidelity is greater than the mean site fidelity for all areas; (3) the area is repeatedly used across years, whereas others are not.

Tien Og is named Shark Bay due to the abundance of sharks throughout the year. However, since we cannot provide a comparison with adjacent areas, we cannot define Tien Og bay as a nursery area. Considering the low mobility of *Carcharhinus melanopterus* juveniles, their constant presence in the bay paired, and the presence of only adult females, we can hypothesize that the area is involved in the reproduction of *Carcharhinus melonapterus*, although the contribution to the adult stock remains unknown. *Carcharhinus melanopterus* may visit the bay during specific times of the day to raise body temperature, potentially facilitating an increase in metabolic rate and/or gestation, and abandon it during the midday hours when temperatures rise excessively. During this study, around 14:00, the water temperature in the bay, as recorded by an Olympus TG-5 camera, reached 31.5 °C.

5. Conclusions

The study encompassed a population of *Carcharhinus melanopterus* in Tien Og Bay on Koh Tao Island, Thailand. Aerial surveys were conducted using a UAV to study the spatio-temporal distribution, abundance, and groups composition of the species. The proposed method has proven to be inexpensive, noninvasive, effective overall, and capable of providing, even in rather short time intervals, valuable information on the ecology of a species that is not always easy to study. In particular, the use of UAVs has allowed us to collect data on several different individuals within defined transects and considering different environmental variables. Further data collection campaigns could be useful to improve the method and the use of UAVs; however, this preliminary study demonstrates how the use of this technology can represent valid support to researchers for the collection of data on the spatial ecology of coastal sharks in shallow waters.

Author Contributions: Conceptualization, A.D.T., G.G. and E.S.; methodology, A.D.T., G.G. and E.S.; software, A.D.T., validation, F.L.L., E.S. and G.G., formal analysis, A.D.T., investigation, A.D.T. and S.S.; resources, A.D.T. and S.S.; data curation, A.D.T.; writing—original draft preparation, A.D.T., E.S. and G.G.; writing—review and editing, F.L.L., E.S. and G.G.; supervision, E.S. and G.G. All authors have read and agreed to the published version of the manuscript.

Funding: This research received no external funding.

Data Availability Statement: Data is contained within the article.

Acknowledgments: The drone, purchased in Thailand, was registered—together with the operator—with the Civil Aviation Authority of Thailand (CAAT) and the National Broadcasting and Telecommunications Commission (NBTC). Insurance was also stipulated to cover accidents up to 1 million baht, as required by law.

Conflicts of Interest: The authors declare no conflict of interest.

References

1. Hobson, E.S. Feeding Behavior in Three Species of Sharks. *Pac. Sci.* **1963**, *17*, 171–194.
2. Stevens, J.D. Life-History and Ecology of Sharks at Aldabra Atoll, Indian Ocean. *Proc. R. Soc. Lond. B.* **1984**, *222*, 79–106. [CrossRef]
3. Sandin, S.A.; Smith, J.E.; DeMartini, E.E.; Dinsdale, E.A.; Donner, S.D.; Friedlander, A.M.; Konotchick, T.; Malay, M.; Maragos, J.E.; Obura, D.; et al. Baselines and Degradation of Coral Reefs in the Northern Line Islands. *PLoS ONE* **2008**, 3, e1548. [CrossRef]
4. Cortes, E. Standardized Diet Compositions and Trophic Levels of Sharks. *ICES J. Mar. Sci.* **1999**, *56*, 707–717. [CrossRef]
5. Last, P.R.; Stevens, J.D. *Sharks and Rays of Australia*; Harvard University Press: Cambridge, MA, USA, 2009; ISBN 978-0-674-03411-2.
6. Chin, A.; Simpfendorfer, C.; Tobin, A.; Heupel, M. Validated Age, Growth and Reproductive Biology of Carcharhinus Melanopterus, a Widely Distributed and Exploited Reef Shark. *Mar. Freshw. Res.* **2013**, *64*, 965. [CrossRef]
7. Lyle, J. Observations on the Biology of *Carcharhinus cautus* (Whitley), *C. melanopterus* (Quoy & Gaimard) and *C. fitzroyensis* (Whitley) from Northern Australia. *Mar. Freshw. Res.* **1987**, *38*, 701. [CrossRef]

8. White, W.T. *Economically Important Sharks & Rays of Indonesia = Hiu dan Pari Yang Bernilai Ekonomis Penting di Indonesia;* Australian Centre for International Agricultural Research: Canberra, Australia, 2006; ISBN 978-1-86320-517-7.
9. Chin, A.; Tobin, A.; Simpfendorfer, C.; Heupel, M. Reef Sharks and Inshore Habitats: Patterns of Occurrence and Implications for Vulnerability. *Mar. Ecol. Prog. Ser.* **2012**, *460*, 115–125. [CrossRef]
10. Mukharror, D.A.; Susiloningtyas, D.; Handayani, T.; Pridina, N. Blacktip Reefshark (Carcharhinus Melanopterus) Individual's Identification in Morotai Waters Using Its Fin's Natural Markings. *AIP Conf. Proc.* **2019**, *2202*, 020085.
11. Lyle, J.M.; Timms, G.J. Predation on Aquatic Snakes by Sharks from Northern Australia. *Copeia* **1987**, *1987*, 802. [CrossRef]
12. Papastamatiou, Y.P.; Lowe, C.G.; Caselle, J.E.; Friedlander, A.M. Scale-dependent Effects of Habitat on Movements and Path Structure of Reef Sharks at a Predator-dominated Atoll. *Ecology* **2009**, *90*, 996–1008. [CrossRef]
13. Mourier, J.; Vercelloni, J.; Planes, S. Evidence of Social Communities in a Spatially Structured Network of a Free-Ranging Shark Species. *Anim. Behav.* **2012**, *83*, 389–401. [CrossRef]
14. Klimley, A.P.; Le Boeuf, B.J.; Cantara, K.M.; Richert, J.E.; Davis, S.F.; Van Sommeran, S.; Kelly, J.T. The Hunting Strategy of White Sharks (Carcharodon Carcharias) near a Seal Colony. *Mar. Biol.* **2001**, *138*, 617–636. [CrossRef]
15. Myrberg, A.A., Jr. Distinctive Markings of Sharks: Ethological Considerations of Visual Function. *J. Exp. Zool.* **1990**, *256*, 156–166. [CrossRef]
16. Butcher, P.A.; Colefax, A.P.; Gorkin, R.A.; Kajiura, S.M.; López, N.A.; Mourier, J.; Purcell, C.R.; Skomal, G.B.; Tucker, J.P.; Walsh, A.J.; et al. The Drone Revolution of Shark Science: A Review. *Drones* **2021**, *5*, 8. [CrossRef]
17. Leonetti, F.L.; Bottaro, M.; Giglio, G.; Sperone, E. Studying Chondrichthyans Using Baited Remote Underwater Video Systems: A Review. *Animals* **2024**, *14*, 1875. [CrossRef] [PubMed]
18. Colefax, A.; Kelaher, B.; Pagendam, D.E.; Butcher, P. Assessing White Shark (Carcharodon Carcharias) Behavior Along Coastal Beaches for Conservation-Focused Shark Mitigation. *Front. Mar. Sci.* **2020**, *7*, 265. [CrossRef]
19. Raoult, V.; Broadhurst, M.; Peddemors, V.; Williamson, J.; Gaston, T. Resource Use of Great Hammerhead Sharks (Sphyrna Mokarran) off Eastern Australia. *J. Fish Biol.* **2019**, *95*, 1430–1440. [CrossRef]
20. Gore, M.; Abels, L.; Wasik, S.; Saddler, L.; Ormond, R. Are Close-Following and Breaching Behaviours by Basking Sharks at Aggregation Sites Related to Courtship? *J. Mar. Biol. Assoc. UK* **2018**, *99*, 693. [CrossRef]
21. Dines, S.; Gennari, E. First Observations of White Sharks (Carcharodon Carcharias) Attacking a Live Humpback Whale (Megaptera Novaeangliae). *Mar. Freshw. Res.* **2020**, *71*, 1205–1210. [CrossRef]
22. Doan, M.; Kajiura, S. Adult Blacktip Sharks (Carcharhinus Limbatus) Use Shallow Water as a Refuge from Great Hammerheads (Sphyrna Mokarran). *J. Fish Biol.* **2020**, *96*, 1530–1533. [CrossRef] [PubMed]
23. Kiszka, J.; Mourier, J.; Gastrich, K.; Heithaus, M. Using Unmanned Aerial Vehicles (UAVs) to Investigate Shark and Ray Densities in a Shallow Coral Lagoon. *Mar. Ecol. Prog. Ser.* **2016**, *560*, 237–242. [CrossRef]
24. Bennett, M.K.; Younes, N.; Joyce, K. Automating Drone Image Processing to Map Coral Reef Substrates Using Google Earth Engine. *Drones* **2020**, *4*, 50. [CrossRef]
25. Yeemin, T.; Sutthacheep, M.; Pettongma, R. Coral Reef Restoration Projects in Thailand. *Ocean Coast. Manag.* **2006**, *49*, 562–575. [CrossRef]
26. Scott, C.M.; Mehrotra, R.; Hein, M.Y.; Moerland, M.S.; Hoeksema, B.W. Population Dynamics of Corallivores (Drupella and Acanthaster) on Coral Reefs of Koh Tao, a Diving Destination in the Gulf of Thailand. *Raffles Bull. Zool.* **2017**, *65*, 68–79.
27. Wongthong, P.; Harvey, N. Integrated Coastal Management and Sustainable Tourism: A Case Study of the Reef-Based SCUBA Dive Industry from Thailand. *Ocean Coast. Manag.* **2014**, *95*, 138–146. [CrossRef]
28. Weterings, R. A GIS-Based Assessment of Threats to the Natural Environment on Koh Tao, Thailand. *Kasetsart J. (Nat. Sci.)* **2011**, *45*, 743–755.
29. Hein, M.Y.; Lamb, J.B.; Scott, C.; Willis, B.L. Assessing Baseline Levels of Coral Health in a Newly Established Marine Protected Area in a Global Scuba Diving Hotspot. *Mar. Environ. Res.* **2015**, *103*, 56–65. [CrossRef]
30. Scaps, P.; Scott, C. An Update to the List of Coral Reef Fishes from Koh Tao, Gulf of Thailand. *Check List* **2014**, *10*, 1123–1133. [CrossRef]
31. Mas, F.; Forselledo, R.; Domingo, A. Length-length relationships for six pelagic shark species commonly caught in the southwestern Atlantic Ocean. *Collect. Vol. Sci. Pap. ICCAT* **2014**, *70*, 2441–2445.

diversity

MDPI

Article

Integrating Egg Case Morphology and DNA Barcoding to Discriminate South American Catsharks, *Schroederichthys bivius* and *S. chilensis* (Carcharhiniformes: Atelomycteridae)

Carlos Bustamante [1,2,*], Carolina Vargas-Caro [1,2], María J. Indurain [1,2,3] and Gabriela Silva [1,2,3]

1 CHALLWA, Laboratorio de Biología Pesquera, Instituto de Ciencias Naturales Alexander von Humboldt, Facultad de Ciencias del Mar y de Recursos Biológicos, Universidad de Antofagasta, Antofagasta 1240000, Chile; c.vargascaro@challwa.org (C.V.-C.); maria.torres.indurain@ua.cl (M.J.I.)

2 Programa de Conservación de Tiburones, Facultad de Ciencias del Mar y de Recursos Biológicos, Universidad de Antofagasta, Antofagasta 1240000, Chile

3 Magíster en Ecología de Sistemas Acuáticos, Universidad de Antofagasta, Antofagasta 1240000, Chile

* Correspondence: carlos.bustamante@uantof.cl; Tel.: +56-55-2637610

Abstract

Catsharks are benthic elasmobranchs that share spatial niches with littoral and demersal bony fishes. The genus *Schroederichthys* includes five species, two of which, *S. chilensis* and *S. bivius*, occur in the waters of Chile. These species are morphologically similar and are often misidentified because of their overlapping external features and color patterns. To improve species discrimination, we analyzed the egg case morphology of both species based on 36 egg cases (12 *S. chilensis*, 24 *S. bivius*) collected from gravid females captured as bycatch in artisanal fisheries between Iquique and Puerto Montt (July–December 2021). Nine morphometric variables were measured and standardized using the total egg case length. Although the egg cases were similar in general appearance, multivariate analyses revealed significant interspecific differences, with egg case height and anterior border width emerging as the most diagnostic variables. Linear discriminant analysis achieved a 100% classification accuracy within this dataset. To confirm species identity, 24 tissue samples (12 per species) were sequenced for the mitochondrial cytochrome c oxidase subunit I (COI) gene. The haplotypes corresponded to previously published sequences from Chile (*S. chilensis*) and Argentina (*S. bivius*), with reciprocal monophyly and 100% bootstrap support. While COI barcoding provided robust confirmation, the core contribution of this study lies in the identification of species-specific egg case morphometrics. Together, these findings establish a dual-track toolkit, egg case morphology for primary discrimination and COI barcodes for confirmatory validation, that can be incorporated into bycatch monitoring and biodiversity assessments, supporting the conservation of poorly known catsharks in the Southeast Pacific.

Keywords: egg capsule; DNA barcoding; pintarroja; oviparous sharks

1. Introduction

Colored catsharks (Carcharhiniformes: Atelomycteridae) are benthic elasmobranchs that share spatial niches with littoral and demersal bony fishes such as soles, hakes, and eels. However, unlike most bony fishes, catsharks possess life-history traits such as low fecundity, late sexual maturity, and relatively high longevity, which render them particularly vulnerable to overfishing [1]. These characteristics present significant conservation challenges [2]. Over the past two decades, global assessments of shark conservation status have highlighted the susceptibility of small-bodied species to threats posed by incidental capture in fisheries [3], with some populations facing serious risks of depletion or local extinction [4].

Within Atelomycteridae, the subfamily Schroederichthyinae comprises six species, five of which are restricted to the Atlantic and Pacific coasts of Central and South America, and the recently described *Akheilos suwartanai* White, Fahmi, Weigmann, 2019, from Indonesia, which extends the subfamily beyond the Americas [5]. The genus *Schroederichthys* includes *S. chilensis* (Guichenot, 1848), distributed from southern Peru to central Chile; *S. bivius* (Smith, 1838), occurring from southern Chile to northern Argentina; *S. maculatus* (Springer, 1966) from the western Atlantic; *S. saurisqualus* (Soto, 2001) from southern Brazil; and *S. tenuis* (Springer, 1966) from northern Brazil to Suriname [6–8]. In Chilean waters, *S. chilensis* and *S. bivius* are thought to occur sympatrically over part of their latitudinal range, although they differ in depth preference: *S. chilensis* occupies subtidal rocky reefs between the Paracas Peninsula (14° S) and Chiloé Island (42° S), whereas *S. bivius* occurs on the upper continental slope between Valdivia (39° S) and the Magallanes region (54° S), and potentially extends east into the southwestern Atlantic [9].

Although catsharks are relatively abundant along the Chilean coast [10,11], there is a lack of biological and taxonomic data on *S. chilensis*, and the available information on *S. bivius* is geographically restricted, with studies focused primarily on Argentina [12,13]. Morphological differentiation between the two species is subtle and based on external features, such as body proportions, nasal flap shape, and dorsal coloration [14]. However, these characteristics are often obscured by phenotypic plasticity, ontogeny, and sexual dimorphism [13], making accurate identification challenging without comparative materials. Springer [14] noted in his global review of catsharks that "it has never been clearly and exhaustively established how to differentiate specimens of *S. chilensis* from *S. bivius*," a taxonomic uncertainty that persists to this day. Early anatomical studies also suggested intrageneric heterogeneity within the genus [6], a view reinforced by White et al. [5] and Soares and Mathubara [15], who argued that *S. bivius* and *S. chilensis* may represent morphologically distinct lineages within the Schroederichthyinae.

Egg case morphology has long been recognized as a valuable character system in elasmobranch taxonomy. These structures frequently exhibit species-specific variations in size, shape, surface features, and pigmentation [16–19] and have been used in a wide range of studies, from reproductive ecology to phylogenetic inference [20–25]. Molecular tools, particularly mitochondrial DNA barcoding of the cytochrome c oxidase subunit I (COI) gene, have emerged as robust complementary methods for species identification [26–29].

For *Schroederichthys chilensis* and *S. bivius*, whose overlapping external morphologies often hinder identification, COI barcoding may provide an independent line of evidence to support species delimitation. When integrated with egg case morphometrics, molecular data can significantly improve taxonomic resolution in ecological and fisheries contexts, particularly in cases where adult specimens are unavailable or damaged. Given the current lack of comparative information on egg case morphology and genetic variation in South American catsharks, this study aimed to evaluate the diagnostic value of egg case morphometrics and mitochondrial COI sequences in distinguishing *S. chilensis* from *S. bivius*. This dual framework may provide a baseline for accurate species identification and offers practical tools for fisheries bycatch monitoring, biodiversity assessment, and conservation planning in the Southeast Pacific.

2. Materials and Methods

A total of 36 egg cases from *Schroederichthys chilensis* ($n = 12$) and *S. bivius* ($n = 24$) were collected from the uteri of 18 individuals incidentally captured as bycatch in artisanal gillnet and longline fisheries along the Chilean coast. Between July and December 2021, specimens of *S. chilensis* were obtained from Iquique (20° S) and Valparaíso (33° S), and *S. bivius* was obtained from Puerto Montt (41° S). For each egg case, the identity of the corresponding female was confirmed using the external diagnostic characters described by Springer [14] and Lamilla and Bustamante [30]. The uterus of origin (left or right) was recorded to test for potential intraspecific morphological differences and rule out any confounding effects on species discrimination. The egg cases were preserved in 70% ethanol for subsequent morphological analysis. In parallel, muscle tissue samples were collected from all gravid females and preserved in 90% ethanol for molecular analysis.

Egg case morphological terminology followed the established methodologies described in the literature [19,21,24]. Nine morphometric variables were recorded for each egg case (Figure 1), including total egg case length (L_{EC}), egg case height (H_{EC}), lateral keel length (LK), anterior border width (W_{AB}), anterior case width (W_{AC}), anterior waist width (W_{AW}), posterior base width (W_{PB}), posterior case width (W_{PC}), and posterior waist width (W_{PW}). Additional measurements of the anterior and posterior length of tendrils (L_{TE}) and the respiratory fissures (RF) were recorded but excluded from comparative analyses because tendrils are prone to natural breakage, which compromises measurement replicability, and the observation of respiratory fissures is highly subjective, limiting the consistency of this variable. All measurements were recorded to the nearest 0.01 mm using a digital caliper and expressed as proportions of L_{EC} to minimize size-related effects. Values are reported as mean $\pm$ standard deviation (s.d.). The spatial orientation of each egg (anterior and dorsal surfaces) was determined based on its position within the uterus, following the criteria described by Gomes and de Carvalho [31]. An independent two-sample t-test was conducted to evaluate potential differences in egg case length (L_{EC}) and posterior case width (W_{PC}) between the species (*S. chilensis* vs. *S. bivius*). To compare the differences in L_{EC} and W_{PC} between the left and right uteri within individual females, a paired t-test was used following confirmation of normality and homogeneity of variances [32].

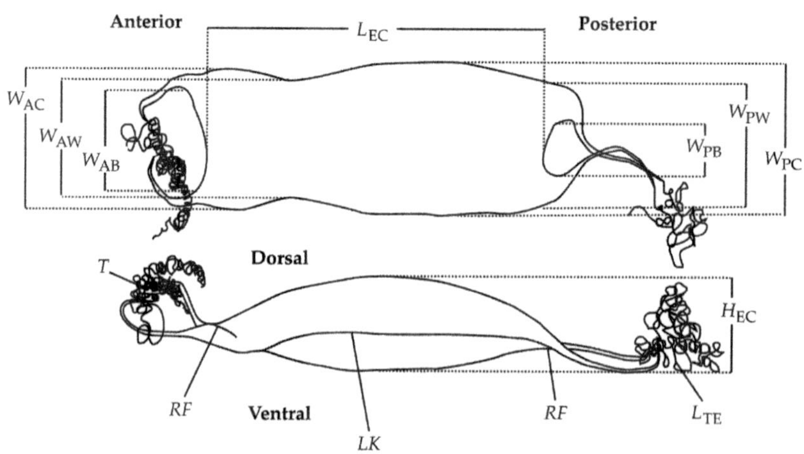

Figure 1. Schematic representation of a catshark-type egg case, showing measurements from the dorsal and lateral views. Abbreviations represent egg case length (L_{EC}), egg case height (H_{EC}), lateral keels (LK), anterior border width (W_{AB}), anterior case width (W_{AC}), anterior waist width (W_{AW}), posterior base width (W_{PB}), posterior case width (W_{PC}) and posterior waist width (W_{PW}), tendrils (T), length of tendrils (L_{TE}) and respiratory fissures (RF). Adapted from Bustamante et al. [24].

Morphometric measurements were used to characterize each egg case, with the identity of the egg-bearing female included as a grouping factor to account for potential non-independence among samples. Analyses were conducted in RStudio (version 2025.09.0) using the MorphoTools2 package (version 1.0.2.1) [33]. The normality of each variable was assessed using the Shapiro–Wilk test, and the homogeneity of variance was tested using Levene's test. Variables were log-transformed where necessary to meet the assumptions of the parametric analysis. To identify and reduce redundancy among variables, Pearson's correlation coefficients were calculated, and highly correlated pairs ($|r| \geq 0.95$) were excluded from analysis. The resulting dataset was used to construct a Euclidean distance matrix for hierarchical clustering and principal component analysis (PCA), allowing for a visual assessment of morphometric similarity among specimens. Linear discriminant analysis (LDA) was performed to assess the discriminatory power of the retained variables [34], generating discriminant functions that maximized the separation between the predefined groups. The assumption of homogeneity of the covariance matrices was evaluated using Box's *M* test [33]. These analyses facilitated the identification of distinct morphotypes associated with species identity. To validate the classification performance, a cross-validation procedure was performed using the *k*-nearest neighbor (*k*-NN) method, which reassigns each specimen to its most likely group based on its morphometric profile and proximity to the group centroids.

Total genomic DNA was extracted from 24 individuals (12 *S. chilensis* and 12 *S. bivius*) using the standard phenol–chloroform protocol [35]. These samples corresponded directly to the gravid females from which egg cases were obtained, establishing a one-to-one correspondence between the morphological and molecular datasets. A fragment of the mitochondrial cytochrome c oxidase subunit I (COI) gene was amplified using the universal primers described by Ward et al. [36], as follows: PCR amplifications were carried out in 10 µL reactions containing 1 µL of genomic DNA (20–50 ng), 5.9 µL of Milli-Q H_2O, 1 µL of 10× buffer with $MgCl_2$ (15 mM), 1 µL of dNTPs (2 mM), 0.5 µL of each primer, and 0.1 µL of Taq DNA polymerase (5 U/µL). The thermal cycling profile consisted of an initial denaturation step at 95 °C for 2 min, followed by 35 cycles of denaturation at 94 °C for 30 s, annealing at 54 °C for 30 s, and extension at 72 °C for 1 min, with a final extension step at 72 °C for 10 min. Amplified PCR products were purified using ExoSAP-IT (USB Products, Affymetrix, Inc., Santa Clara, CA, USA) by incubation at 37 °C for 45 min, followed by enzyme inactivation at 80 °C for 15 min. Sequencing was performed bidirectionally at the Austral-Omics Sequencing Facility (Valdivia, Chile). Sequence chromatograms were manually checked for base-calling errors, and variable sites were verified by comparing forward and reverse reads. Primer sequences were trimmed, and consensus sequences were assembled for each individual. Intraspecific genetic distances (*p*) were estimated using pairwise comparisons under the Kimura 2-parameter (K2P) model [37] in MEGA v7 [38]. The K2P distance matrix was also used to construct a neighbor-joining (NJ) tree in GENEIOUS [39], with node support assessed using 1000 bootstrap replicates. Novel haplotypes were deposited in the NCBI GenBank database under the accession numbers PX250298–PX250321. While GenBank remains a repository for homologous sequences of *Schroederichthys*, we acknowledge the risk of misidentified entries and therefore relied on previously validated haplotypes from Chile and Argentina to ensure confidence in species identity. The final alignments for COI included additional Scyliorhinidae species as outgroups to provide a phylogenetic framework for species discrimination (Table 1).

Table 1. List of colored catsharks (Atelomycteridae) and related species within the Carcharhiniformes (including the outgroup) used in this study. The COI sequence accession numbers from the NCBI GenBank database are indicated. Novel sequences are indicated in bold.

Species	Locality	Accession Number
Carcharhiniformes: Atelomycteridae		
Atelomycterus baliensis	Indonesia: Bali	EU398568, EU398569
Atelomycterus erdmanni	Indonesia	KP769787
Atelomycterus fasciatus	Australia: Western Australia	EU398570
Atelomycterus marmoratus	Philippines: Cebu	OQ386292, OQ386496
Atelomycterus marnkalha	Australia: Queensland	EU398574, EU398577
Aulohalaelurus labiosus	Australia: Western Australia	EU398581, HQ955988, JN312813
Schroederichthys bivius	Argentina	EU074581–EU074586
Schroederichthys bivius	Chile: Valdivia	**PX250298–PX250303**
Schroederichthys bivius	Chile: Puerto Montt	**PX250304–PX250309**
Schroederichthys chilensis	Chile: Coquimbo	MK982882, MK982902
Schroederichthys chilensis	Chile: Iquique	**PX250310, PX250311**
Schroederichthys chilensis	Chile: Valparaíso	**PX250312–PX250315**
Schroederichthys chilensis	Chile: San Antonio	**PX250316–PX250321**
Carcharhiniformes: Scyliorhinidae		
Cephaloscyllium laticeps	Australia: Victoria	HQ956280
Cephaloscyllium pictum	Indonesia: Nusa Tenggara Barat	EU398676
Cephaloscyllium silasi	India	KF899711
Cephaloscyllium variegatum	Australia: Tasmania	HQ956282
Poroderma africanum	South Africa: Western Cape	OR138385, OR138381
Poroderma pantherinum	South Africa: Western Cape	OR138386, OR138391
Scyliorhinus canicula	USA: Washington	KJ709897, JN641232
Scyliorhinus capensis	South Africa: Western Cape	OR138392
Scyliorhinus stellaris	Malta	KJ709900
Carcharhiniformes: Pentanchidae		
Apristurus brunneus	USA: Washington	JQ353982
Apristurus melanoasper	Canada: Newfoundland	MW339366
Apristurus nasutus	Chile: Coquimbo	KU737638
Apristurus profundorum	Canada: Newfoundland	MW339382
Holohalaelurus punctatus	South Africa	OR138367, OR138368
Holohalaelurus regani	South Africa: Western Cape	OR138372, OR138373
Asymbolus analis	Australia	HM902609
Asymbolus parvus	Australia: Western Australia	EU398564
Asymbolus rubiginosus	Australia	EU398567
Carcharhiniformes: Carcharhinidae		
Carcharhinus brachyurus [1]	South Korea: Jeju island	MT995631

[1] Species included as an outgroup.

3. Results

3.1. Description of the Egg Cases

The egg cases of *Schroederichthys chilensis* (Figure 2a,b; Table 2) were relatively small, with a pronounced anterior waist and no horns or tendrils at the anterior end of the egg case. The surface was smooth on both the dorsal and ventral faces and was covered with a thin layer of fine filaments. The overall color was light brown. Near the anterior respiratory fissures, a bundle of fibrils originated that matched the thickness of the posterior horns. Posteriorly, the egg cases exhibited two well-developed horns, each giving rise to a long, robust tendril. These tendrils gradually tapered, reaching approximately one-third of their original diameter at the distal end of the tendril.

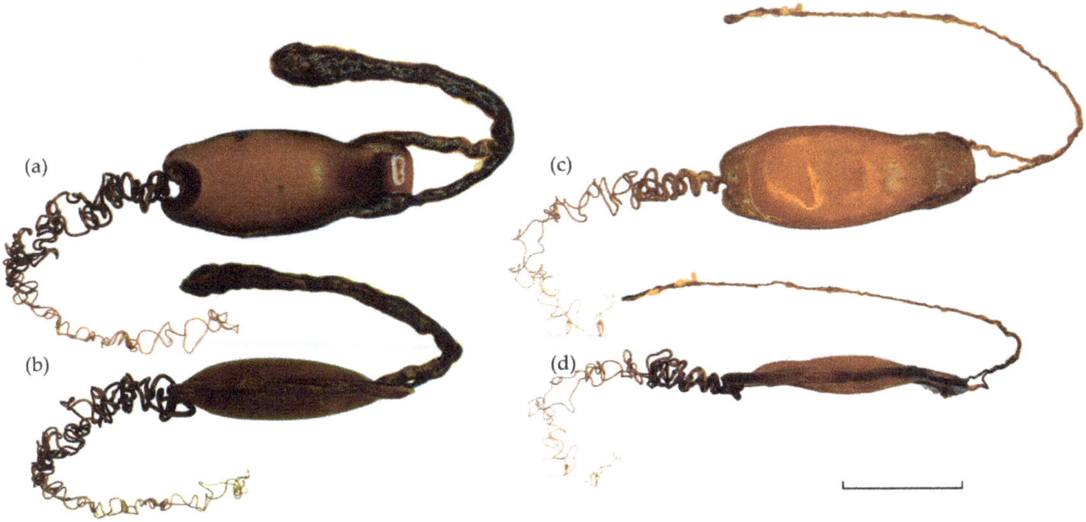

Figure 2. Egg cases of Chilean catsharks: (**a**,**b**) *Schroederichthys chilensis* and (**c**,**d**) *S. bivius* in the dorsal (**a**,**c**) and (**b**,**d**) lateral views. The scale bar represents a length of 20 mm.

Table 2. Measurements of the egg cases of *Schroederichthys chilensis* and *S. bivius* from Chilean waters. The mean values and standard deviations are expressed as percentages of L_{EC}. The abbreviations used for each measurement are presented in Figure 1.

Measurement	*Schroederichthys chilensis*		*Schroederichthys bivius*	
	Mean (s.d.)	Range (mm)	Mean (s.d.)	Range (mm)
L_{EC}	52.3 (3.8)	45.2–57.1	52.8 (3.2)	45.0–58.6
H_{EC}	11.0 (1.8)	7.8–13.7	4.9 (1.6)	2.5–10.0
LK	3.3 (0.4)	2.8–4.0	2.0 (0.9)	0.1–3.7
W_{AB}	14.0 (0.4)	13.3–14.6	9.1 (1.3)	7.2–13.4
W_{AC}	17.4 (0.9)	15.7–18.5	13.0 (1.5)	10.2–16.6
W_{AW}	16.9 (1.1)	14.9–18.2	14.9 (1.1)	13.0–18.1
W_{PB}	5.8 (1.3)	4.1–7.7	6.7 (1.8)	3.2–9.2
W_{PC}	24.2 (1.7)	21.2–26.2	19.7 (2.3)	12.2–23.8
W_{PW}	18.7 (1.5)	16.1–20.3	13.7 (2.0)	10.7–18.1

The egg cases of *S. bivius* (Figure 2c,d; Table 2) were similar in general shape and size to those of *S. chilensis* but had several distinguishing features. Like *S. chilensis*, the anterior end of *S. bivius* egg cases lacked horns and tendrils; however, they displayed a characteristic double waist at the anterior margin. The color was light brown, and the surface texture was smooth. Fine filaments formed elongated thread-like cords near the anterior respiratory fissures. The posterior horns gave rise to tendrils that narrowed abruptly, tapering to approximately one-third of the egg case length before terminating in delicate filamentous tips.

3.2. Morphometric Analyses

All morphometric variables met the assumption of normality, as assessed using the Shapiro–Wilk test ($p > 0.05$). Within each species, no significant differences were detected in egg case length (L_{EC}) or posterior case width (W_{PC}) between the left and right uteri, indicating the absence of intra-individual asymmetry in these measurements. For *S. bivius*, paired *t*-tests showed no significant differences in L_{EC} ($t = -0.0916$, $p = 0.8872$) or W_{PC} ($t = -1.0162$, $p = 0.3313$), and for *S. chilensis*, the results were similarly non-significant (L_{EC}:

t = −0.6094, p = 0.8125; W_{PC}: t = 1.4363, p = 0.2104). Interspecific comparisons revealed a significant difference in posterior case width, with *S. chilensis* exhibiting broader egg cases than *S. bivius* (t = 5.952, p < 0.001). No significant difference was observed in the egg case length between species (t = 0.5223, p = 0.604). All morphometric variables met the assumptions of normality (Shapiro–Wilk test, p > 0.05) and homogeneity of variance (Levene's test, p > 0.05). Following correlation filtering, eight variables were retained for multivariate analyses. Hierarchical clustering of Euclidean distance values revealed two non-overlapping groups corresponding to the two species (Figure 3). This strong species-level separation was consistent across individuals. The mean intraspecific dissimilarity was low (1.1–1.3 standardized Euclidean units), whereas interspecific dissimilarity was comparatively high (4.9 standardized Euclidean units), confirming a clear morphometric divergence between the two species.

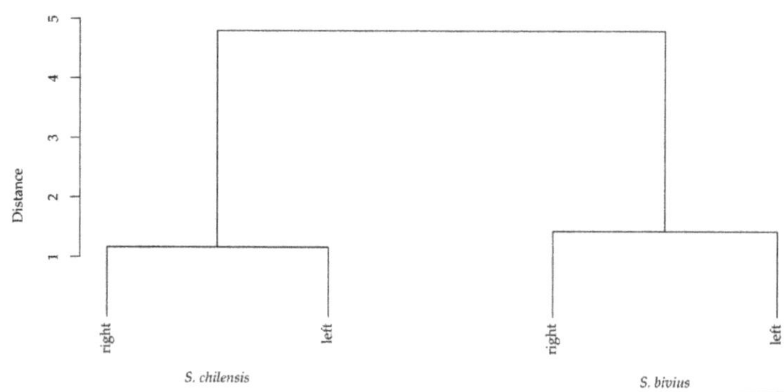

Figure 3. Comparative analysis of the left and right egg cases (associated with the uteri) of Chilean catsharks (*Schroederichthys chilensis* and *S. bivius*) based on the Euclidean distance of eight morphometric variables standardized by the total egg case length.

The PCA ordination revealed a clear separation between *S. bivius* and *S. chilensis* along the first two principal components, with PC1 and PC2 explaining 67.84% and 12.88% of the total variance, respectively (Figure 4a). The eigenvector biplot (Figure 4b) shows that the most influential variables along PC1 included anterior case width (W_{AC}), lateral keel length (LK), egg case height (H_{EC}), anterior border width (W_{AB}), and posterior case width (W_{PC}), all of which contributed to the species-level differentiation. Linear discriminant analysis (LDA) further supported species discrimination. Box's M-test indicated homogeneity of covariance matrices (χ^2 = 2.72, df = 3, p = 0.437). Two variables, W_{AB} and H_{EC}, significantly contributed to group separation. The classification model correctly assigned all egg cases to their respective species, achieving 100% classification accuracy within this dataset. All *S. bivius* (n = 24) and *S. chilensis* (n = 12) egg cases were accurately classified based on the retained morphometric variables (Table 3). The k-nearest neighbor classification analysis showed that the highest number of correct classifications was at k = 3 and reflected the LDA results.

Table 3. Species classifications ('as.', e.g., as. *S. chilensis*) of Chilean catsharks (*Schroederichthys chilensis and S. bivius*) by using a linear discriminant analysis (LDA) based on eight egg case morphometric measurements.

Taxon	n	as. *S. chilensis*	as. *S. bivius*	n Correct	Percentage Correct
S. chilensis	12	12	0	12	100
S. bivius	24	0	24	24	100
Total	36	12	24	36	100

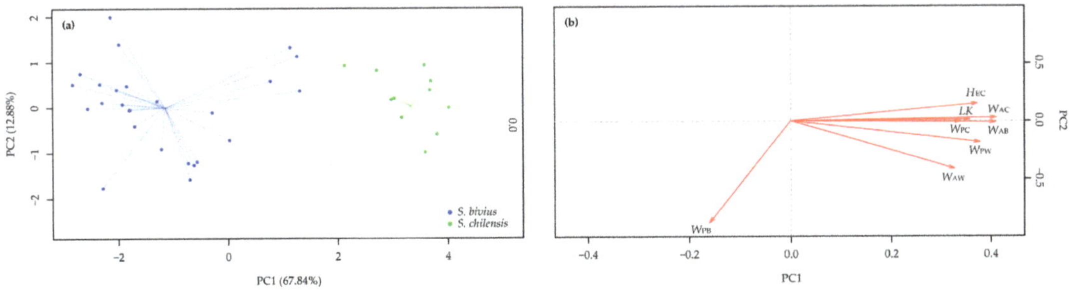

Figure 4. Principal component analysis (**a**) and eigenvector biplot (**b**) of the morphometric analysis of egg cases from Chilean catsharks (*Schroederichthys chilensis and S. bivius*) based on the Euclidean distance of eight morphometric variables standardized by the total egg case length.

3.3. Molecular Analyses

A total of 24 high-quality COI sequences were obtained (12 *S. chilensis*, 12 *S. bivius*), corresponding directly to the gravid females from which egg cases were analyzed. The sequence lengths ranged from 685 to 763 bp after trimming. No insertions, deletions, or stop codons were observed. Intraspecific K2P genetic distances were low (0.0021 for *S. chilensis*, 0.0018 for *S. bivius*), whereas interspecific divergence was 0.053, consistent with species-level separation in elasmobranchs. Two haplotypes were detected in *S. bivius* and one in *S. chilensis*, all of which matched previously published sequences from Argentina and Chile, respectively. Phylogenetic comparisons were performed using 67 sequences and 30 closely related species. The neighbor-joining tree based on K2P distances showed two well-supported clades corresponding to *S. chilensis* and *S. bivius*, with 100% bootstrap support (Figure 5). No haplotypes were shared among the species. All sequences were grouped with reference sequences, confirming the correct species assignment and reflecting the current taxonomy of the Atelomycteridae family.

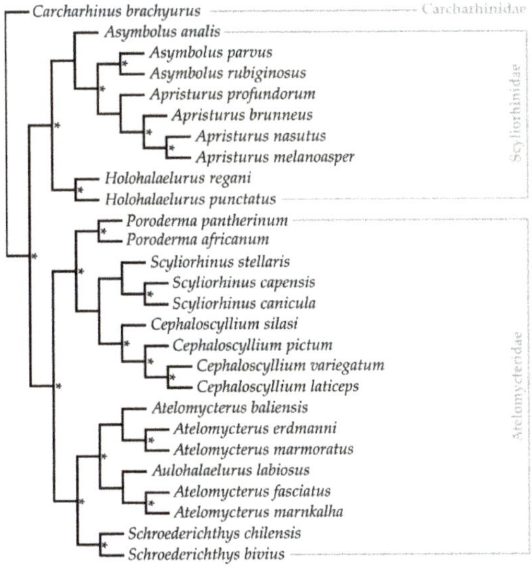

Figure 5. Phylogenetic tree of Chilean catsharks (Carcharhiniformes: Atelomycteridae) inferred from novel and reference COI sequences using a Neighbor-Joining (NJ) tree based on K2P parameter distances. The species *Carcharhinus brachyurus* was used as an outgroup representative to root the tree. Bootstrap support at branch nodes is shown (with a *) when > 80%. Families are represented outside the tip labels of the tree.

4. Discussion

This study provides an integrative approach to distinguish between two morphologically similar catshark species, *Schroederichthys bivius* and *S. chilensis*, using a combination of egg case morphometrics and mitochondrial DNA barcoding. Although both species produce externally similar egg cases, quantitative analyses revealed consistent species-specific differences, with egg case height and anterior border width emerging as robust diagnostic traits for species identification. The strong discriminatory signal was confirmed by linear discriminant analysis, which achieved 100% classification accuracy within the dataset. These findings validate the strong diagnostic power of reproductive characteristics and highlight the utility of egg case morphology as a practical tool for taxonomic identification in the absence of adults in the population. While this study did not explore environmental drivers, it aligns with previous reports [40], which suggested that the tendril structure of *Schroederichthys* egg cases enhances attachment to the structurally complex kelps, such as *Lessonia trabeculata*. This substrate specificity likely exerts selective pressure on stable morphological traits, such as posterior tendrils and consistent egg case shape, supporting their use as reliable taxonomic markers. The observed interspecific differences in egg case morphometrics appear to be robust across various environmental conditions, reinforcing the value of these characteristics for species-level identification.

The COI barcoding provided independent confirmation of species identity, recovering two reciprocally monophyletic clades with strong bootstrap support and no haplotype sharing. Intraspecific divergence was low, whereas interspecific divergence (5.3%) was well within the range typically observed among closely related elasmobranch taxa [26–29]. The neighbor-joining phylogeny showed reciprocal monophyly with 100% bootstrap support and no shared haplotypes, supporting previous studies that used COI as a robust marker for species identification in elasmobranchs [41,42]. These findings reinforce the value of DNA barcoding as a confirmatory tool that reinforces morphological findings, particularly in taxonomically challenging groups such as scyliorhinids [43–45]. In this study, COI barcoding was applied as a confirmatory marker because it remains the most widely applied and standardized barcode in vertebrate taxonomy. Although single-gene barcoding is limited in its phylogenetic resolution, it provides a robust baseline for species-level validation. Future studies should expand to multilocus or genomic approaches (e.g., NADH2, SNPs, RADseq) to test species boundaries across broader ranges and potential contact zones.

Egg case morphology has played an important role in elasmobranch taxonomy, and our findings add to the growing body of work that recognizes its value beyond reproductive biology. Previous studies have shown that the morphology of egg cases reflects both phylogenetic constraints and ecological adaptations [18,21]. Our comparative analysis confirmed the presence of stable, species-specific egg case features in South American catsharks and supports their continued use in systematic, evolutionary, and field-based identification studies of this group [43]. In practice, these morphometric characters may provide a rapid and non-lethal method for identifying species in fisheries bycatch, biodiversity surveys, and museum collections. When uncertainty persists, COI barcoding can be applied as a confirmatory layer, offering a straightforward workflow that integrates field identification and molecular verification. Such dual-track approaches can be readily adopted by fishery observer programs and national biodiversity monitoring initiatives, ensuring accurate species records even in the absence of adult specimens. Linking egg case occurrence with habitat information, such as the role of kelp forests as keystone habitats, further connects taxonomy to ecosystem-based management and conservation planning.

From a taxonomic perspective, our results are consistent with recent revisions of catshark systematics. Historically placed within Scyliorhinidae, recent systematic revisions have recognized the family Atelomycteridae, with *Schroederichthyinae* as one of its subfamilies [5]. The description of *A. suwartanai* from Indonesia extends the distribution of this subfamily beyond South America, underscoring its previously underestimated diversity. Earlier anatomical studies also suggested that *S. bivius* and *S. chilensis* differ substantially from the "neotenic group" (*S. maculatus*, *S. saurisqualus*, and *S. tenuis*), raising the possibility that Chilean–Patagonian species represent a distinct lineage within the group [6,45]. Recently, Soares and Mathubara [15], using a robust phylogenetic framework based on 143 morphological traits and NADH2 gene sequences, confirmed the placement of *Schroederichthys* within Atelomycteridae and demonstrated that *S. bivius* and *S. chilensis* formed a well-supported monophyletic clade distinct from other scyliorhinids. Our results reinforce this updated classification, providing additional evidence from egg case morphology and mitochondrial COI data, highlighting the evolutionary distinctiveness of these species. The diagnostic egg case traits identified here are consistent with their separation from traditional Scyliorhinidae and support the monophyly of *Schroederichthys* within the Atelomycteridae.

Geographically, *S. bivius* and *S. chilensis* span broad latitudinal and bathymetric ranges along the southeastern Pacific, with a likely overlap between approximately 39°S (Valdivia) and 42°S (Chiloé Island). While *S. chilensis* is typically found in shallower subtidal zones, *S. bivius* occupies deeper shelves and upper continental slope habitats. However, field identification of either species remains difficult due to overlapping external features and high phenotypic plasticity in coloration and body proportions. Springer [14] originally distinguished *S. bivius* from *S. chilensis* by its pointed snout, narrower head, larger eyes, and longer nasal flaps. However, the diagnosis of these species remains incomplete and has not been comprehensively updated using modern morphometric or molecular tools.

Despite these advances, several limitations must be considered. Our sample sizes were modest, and the geographic coverage was restricted to Chilean waters. Future research should expand sampling across the full distributional range of both species, particularly in zones of potential sympatry. Additional studies should explore ontogenetic variation in egg-case morphology and incorporate environmental drivers (e.g., temperature and substrate) that may influence phenotypic traits. From a genetic perspective, multilocus or genome-wide approaches would strengthen species delimitation and test hypotheses, such as cryptic diversity or hybridization in contact zones.

Although *S. bivius* and *S. chilensis* are not currently the targets of commercial fisheries, they are often caught as bycatch in both artisanal and industrial fisheries along the Chilean coast [10]. However, no quantitative estimates of fishing mortality have been reported [11], leaving the impact on their populations unassessed. Given the low reproductive output and life-history traits of catsharks, such as late maturity, low fecundity, and long lifespan, unmonitored bycatch poses a clear conservation risk to these species [46,47]. These risks are exacerbated in coastal and demersal ecosystems, where habitat degradation compounds fishing mortality. Kelp forests, which dominate many subtidal zones along the Chilean coast, act as critical nursery and refuge habitats for diverse fish assemblages, enhancing the survival and recruitment of early life stages [48,49]. Therefore, the loss or fragmentation of these foundational habitats may indirectly affect the persistence of small benthic elasmobranchs, including *Schroederichthys* species, by altering prey availability and reducing shelter from predators [50]. In this context, the integrative framework developed here provides operational tools for monitoring and conservation efforts. Morphometric analysis of egg cases offers a rapid and non-invasive method for identifying species in the field, whereas COI barcoding provides a confirmatory layer when morphology alone is inconclusive. Together, these approaches form a dual-track identification workflow that can be directly incorporated into bycatch observer programs, biodiversity surveys and museum collections. Linking species-specific egg case occurrence with habitat information, such as the role of kelp forests as nurseries, connects species-level taxonomy with ecosystem-based management. As human activities intensify pressures on coastal ecosystems, adopting such integrated identification tools will be essential to strengthen biodiversity assessments, mitigate bycatch impacts, and guide effective conservation of South American catsharks in the Southeast Pacific.

Author Contributions: Conceptualization, C.B.; methodology, C.B. and C.V.-C.; formal analysis, M.J.I. and G.S.; writing—original draft preparation, C.V.-C., M.J.I. and G.S.; writing—review and editing, C.B.; funding acquisition, C.B. All authors have read and agreed to the published version of the manuscript.

Funding: This research was funded by Agencia Nacional de Investigación y Desarrollo de Chile (ANID FONDECYT), grant number 11220358.

Institutional Review Board Statement: This study did not involve animal experimentation or harm. Specimens were obtained as bycatch from commercial fisheries, according to Chilean Law.

Data Availability Statement: The datasets generated and/or analyzed during the current study are contained within the article.

Acknowledgments: The authors would like to acknowledge T. Valencia, who contributed to the initial sampling stage of this project. Special thanks to I. Contreras, F. Concha, R. Oliva, R. Pedrero, F. Palacios, K. Górski, and to the staff of "Programa de Conservación de Tiburones", who aided specimen sorting and sampling.

Conflicts of Interest: The authors declare no conflicts of interest.

References

1. Dulvy, N.K.; Pacoureau, N.; Rigby, C.L.; Pollom, R.A.; Jabado, R.W.; Ebert, D.A.; Finucci, B.; Pollock, C.M.; Cheok, J.; Derrick, D.H.; et al. Overfishing drives over one-third of all sharks and rays toward a global extinction crisis. *Curr. Biol.* **2021**, *31*, 4773–4787.e8. [CrossRef]
2. Pacoureau, N.; Carlson, J.K.; Kindsvater, H.K.; Rigby, C.L.; Winker, H.; Simpfendorfer, C.A.; Charvet, P.; Pollom, R.A.; Barreto, R.; Sherman, C.S.; et al. Conservation successes and challenges for wide-ranging sharks and rays. *Proc. Natl. Acad. Sci. USA* **2023**, *120*, e2216891120. [CrossRef] [PubMed]
3. Sherman, C.S.; Simpfendorfer, C.A.; Pacoureau, N.; Matsushiba, J.H.; Yan, H.F.; Walls, R.H.L.; Rigby, C.L.; VanderWright, W.J.; Jabado, R.W.; Pollom, R.A.; et al. Half a century of rising extinction risk of coral reef sharks and rays. *Nat. Commun.* **2023**, *14*, 15. [CrossRef] [PubMed]
4. Dulvy, N.K.; Pacoureau, N.; Matsushiba, J.H.; Yan, H.F.; VanderWright, W.J.; Rigby, C.L.; Finucci, B.; Sherman, C.S.; Jabado, R.W.; Carlson, J.K.; et al. Ecological erosion and expanding extinction risk of sharks and rays. *Science* **2024**, *386*, eadn1477. [CrossRef] [PubMed]
5. White, W.T.; Fahmi; Weigmann, S. A new genus and species of catshark (Carcharhiniformes: Scyliorhinidae) from eastern Indonesia. *Zootaxa* **2019**, *4691*, 444–460. [CrossRef]
6. Gomes, U.L.; Peters, G.O.; De Carvalho, M.R.; Gadig, O.B.F. Anatomical investigation of the slender catshark *Schroederichthys tenuis* Springer, 1966, with notes on intrageneric relationships (Chondrichthyes: Carcharhiniformes: Scyliorhinidae). *Zootaxa* **2006**, *1119*, 29–58. [CrossRef]
7. Weigmann, S. Annotated checklist of the living sharks, batoids and chimaeras (Chondrichthyes) of the world, with a focus on biogeographical diversity. *J. Fish Biol.* **2016**, *88*, 837–1037. [CrossRef]
8. Ebert, D.A.; Dando, M.; Fowler, S. *Sharks of the World. A Complete Guide*; Princeton University Press: Princeton, NJ, USA, 2021; pp. 429–433.
9. Bornatowski, H.; Santos, L.; Robert, M.D.C.; Weiser, P.A. Occurrence of the narrowmouth catshark *Schroederichthys bivius* (Chondrichthyes: Scyliorhinidae) in southern Brazil. *Mar. Biodivers. Rec.* **2014**, *7*, e51. [CrossRef]
10. Lamilla, J.; Bustamante, C.; Roa, R.; Acuña, E.; Vargas-Caro, C. *Estimación del Descarte de Condrictios en Pesquerías Artesanales*; Universidad Austral de Chile: Valdivia, Chile, 2010.
11. Bustamante, C.; Acuña, E.; Tapia-Jopia, C.; Vargas-Caro, C. *Actualización del Plan de Acción Nacional Para la Conservación y Manejo de Tiburones de Chile*; Universidad de Antofagasta: Antofagasta, Chile, 2023; pp. 211–232.

12. Sánchez, F.; Marí, N.R.; Bernardele, J.C. Distribución, abundancia relativa y alimentación de pintarroja *Schroederichthys bivius* Müller & Henle, 1838 en el Océano Atlántico sudoccidental. *Rev. Biol. Mar. Oceanogr.* **2009**, *44*, 453–466. [CrossRef]
13. Colonello, J.H.; Cortés, F.; Belleggia, M. Male-biased sexual size dimorphism in sharks: The narrowmouth catshark *Schroederichthys bivius* as case study. *Hydrobiologia* **2020**, *847*, 1873–1886. [CrossRef]
14. Springer, S. A revision of the catsharks, family Scyliorhinidae. *NOAA Tech. Rep. NMFS Circ.* **1979**, *422*, 1–153.
15. Soares, K.D.A.; Mathubara, K. Combined phylogeny and new classification of catsharks (Chondrichthyes: Elasmobranchii: Carcharhiniformes). *Zool. J. Linn. Soc.* **2022**, *195*, 761–814. [CrossRef]
16. Ishiyama, R. Studies on the rays and skates belonging to the family Rajidae, found in Japan and adjacent regions. 1 Egg-capsule of ten species. *Jap. J. Ichthyol.* **1950**, *1*, 30–36.
17. Nakaya, K. Taxonomy, comparative anatomy and phylogeny of Japanese catsharks, Scyliorhinidae. *Mem. Fac. Fish. Hokkaido Univ.* **1975**, *23*, 1–94.
18. Musick, J.A.; Ellis, J.K. Reproductive evolution of chondrichthyans. In *Reproductive Biology and Phylogeny of Chondrichthyes: Sharks, Rays and Chimaeras*; Hamlett, W.C., Ed.; Science Publishers: Enfield, NH, USA, 2005; pp. 45–79.
19. Ebert, D.A.; Compagno, L.J.V.; Cowley, P.D. Reproductive biology of catsharks (Chondrichthyes: Scyliorhinidae) off the west coast of southern Africa. *ICES J. Mar. Sci.* **2006**, *63*, 1053–1065. [CrossRef]
20. Dulvy, N.K.; Reynolds, J.D. Evolutionary transitions among egg-laying, live-bearing and maternal inputs in sharks and rays. *Proc. R. Soc. Lond. B* **1997**, *264*, 1309–1315. [CrossRef]
21. Flammang, B.E.; Ebert, D.A.; Cailliet, G.M. Egg cases of the genus *Apristurus* (Chondrichthyes: Scyliorhinidae): Phylogenetic and ecological implications. *Zoology* **2007**, *110*, 308–317. [CrossRef]
22. Kyne, P.M.; Simpfendorfer, C.A. Deepwater Chondrichthyans. In *Biology of Sharks and Their Relatives II: Biodiversity, Adaptive Physiology, and Conservation*; Carrier, J.C., Musick, J.A., Heithaus, M.R., Eds.; CRC Press: Boca Raton, FL, USA, 2010; pp. 37–114.
23. Mabragaña, E.; Figueroa, D.; Scenna, L.; Díaz de Astarloa, J.; Colonello, J.; Delpiani, G. Chondrichthyan egg cases from the south-west Atlantic Ocean. *J. Fish Biol.* **2011**, *79*, 1261–1290. [CrossRef]
24. Bustamante, C.; Kyne, P.M.; Bennett, M.B. Comparative morphology of the egg cases of *Asymbolus* analis, *Asymbolus rubiginosus* and *Figaro boardmani* (Carcharhiniformes: Scyliorhinidae) from southern Queensland, Australia. *J. Fish Biol.* **2013**, *83*, 133–143. [CrossRef]
25. Mancusi, C.; Massi, D.; Baino, R.; Cariani, A.; Crobe, V.; Ebert, D.A.; Ferrari, A.; Gordon, C.A.; Hoff, G.R.; Iglesias, S.P.; et al. An identification key for Chondrichthyes egg cases of the Mediterranean and Black Sea. *Eur. Zool. J.* **2021**, *88*, 436–448. [CrossRef]
26. Hebert, P.D.N.; Cywinska, A.; Ball, S.L.; deWaard, J.R. Biological identifications through DNA barcodes. *Proc. Biol. Sci.* **2003**, *270*, 313–321. [CrossRef] [PubMed]
27. Holmes, B.H.; Steinke, D.; Ward, R.W. Identification of shark and ray fins using DNA barcoding. *Fish. Res.* **2009**, *95*, 280–288. [CrossRef]
28. Naylor, G.J.P.; Caira, J.N.; Jensen, K.; Rosana, K.A.M.; Straube, N.; Lakner, C. Elasmobranch Phylogeny: A Mitochondrial Estimate Based on 595 Species. In *Biology of Sharks and Their Relatives*, 2nd ed.; Carrier, J.C., Musick, J.A., Heithaus, M.R., Eds.; CRC Press: Boca Raton, FL, USA, 2012; pp. 31–56.
29. Veríssimo, A.; Zaera-Perez, D.; Leslie, R.; Iglésias, S.P.; Séret, B.; Grigoriou, P.; Sterioti, A.; Gubili, C.; Barría, C.; Duffy, C.; et al. Molecular diversity and distribution of eastern Atlantic and Mediterranean dogfishes *Squalus* highlight taxonomic issues in the genus. 2017. *Zool. Scr.* **2017**, *46*, 414–428. [CrossRef]
30. Lamilla, J.; Bustamante, C. *Guía Para el Reconocimiento de: Tiburones, Rayas y Quimeras de Chile*; Oceana: Santiago, Chile, 2005; pp. 31–32.
31. Gomes, U.L.; de Carvalho, M.R. Egg capsules of *Schroederichthys tenuis* and *Scyliorhinus haeckelii* (Chondrichthyes, Scyliorhinidae). *Copeia* **1995**, *1995*, 232–236. [CrossRef]
32. Fay, M.P.; Proschan, M.A. Wilcoxon-Mann-Whitney or t-test? On assumptions for hypothesis tests and multiple interpretations of decision rules. *Stat. Surv.* **2010**, *4*, 1–39. [CrossRef]
33. Šlenker, M.; Koutecký, P.; Marhold, K. MorphoTools2: An R package for multivariate morphometric analysis. *Bioinformatics* **2022**, *38*, 2954–2955. [CrossRef]
34. Venables, W.N.; Ripley, B.D. Exploratory Multivariate Analysis. In *Modern Applied Statistics with S*, 4th ed.; Venables, W.N., Ripley, B.D., Eds.; Springer: New York, NY, USA, 2002; pp. 301–330.
35. Sambrook, J.; Russell, D.W. *Molecular Cloning: A Laboratory Manual*, 3rd ed.; Cold Spring Harbor Laboratory Press: New York, NY, USA, 2001.
36. Ward, R.D.; Zemlak, T.S.; Innes, B.H.; Last, P.R.; Hebert, P.D. DNA barcoding Australia's fish species. *Philos. Trans. R. Soc. Lond. B Biol. Sci.* **2005**, *360*, 1847–1857. [CrossRef]
37. Kimura, M. A simple method for estimating evolutionary Rrte of base substitutions through comparative studies of nucleotide sequences. *J. Mol. Evol.* **1980**, *16*, 111–120. [CrossRef]

38. Kumar, S.; Stecher, G.; Tamura, K. MEGA7: Molecular evolutionary genetics analysis version 7.0 for bigger datasets. *Mol. Biol. Evol.* **2016**, *33*, 1870–1874. [CrossRef]
39. Kearse, M.; Moir, R.; Wilson, A.; Stones-Havas, S.; Cheung, M.; Sturrock, S.; Buxton, S.; Cooper, A.; Markowitz, S.; Duran, C.; et al. Geneious Basic: An integrated and extendable desktop software platform for the organization and analysis of sequence data. *Bioinformatics* **2012**, *28*, 1647–1649. [CrossRef]
40. Trujillo, J.; Pardo, L.M.; Vargas-Chacoff, L.; Valdivia, N. Sharks in the forest: Relationships between kelp physical-complexity attributes and egg deposition sites of the red-spotted catshark. *Mar. Ecol. Prog. Ser.* **2018**, *610*, 125–135. [CrossRef]
41. Cardeñosa, D.; Fields, A.T.; Shea, S.K.H.; Feldheim, K.A.; Chapman, D.D. Relative contribution to the shark fin trade of Indo-Pacific and Eastern Pacific pelagic thresher sharks. *Anim. Conserv.* **2021**, *24*, 367–372. [CrossRef]
42. Domingues, R.R.; Bunholi, I.V.; Pinhal, D.; Antunes, A.; Mendonça, F.F. From molecule to conservation: DNA-based methods to overcome frontiers in the shark and ray fin trade. *Conserv. Genet. Resour.* **2021**, *13*, 231–247. [CrossRef]
43. Soares, K.D.A.; de Carvalho, M.R. The catshark genus *Scyliorhinus* (Chondrichthyes: Carcharhiniformes: Scyliorhinidae): Taxonomy, morphology and distribution. *Zootaxa* **2019**, *4601*, 1–147. [CrossRef]
44. Soares, K.D.A.; de Carvalho, M.R. Phylogenetic relationship of catshark species of the genus *Scyliorhinus* (Chondrichthyes, Carcharhiniformes, Scyliorhinidae) based on comparative morphology. *Zoosyst. Evol.* **2020**, *96*, 345–395. [CrossRef]
45. Soares, K.D.A.; Zanini, F. Redescription and anatomical investigation of *Schroederichthys maculatus* Springer, 1966 and *S. saurisqualus* Soto, 2001 with comments on their systematics (Chondrichthyes: Carcharhiniformes: Atelomycteridae). *Zool. Anz.* **2023**, *302*, 224–238. [CrossRef]
46. Dulvy, N.K.; Acuña, E.; Bustamante, C.; Chiaramonte, G.E.; Cuevas, J.M.; Herman, K.; Pompert, J.; Velez-Zuazo, X. Schroederichthys bivius. In *The IUCN Red List of Threatened Species*; IUCN: Cambridge, UK, 2020; p. e.T39347A2906921. [CrossRef]
47. Dulvy, N.K.; Acuña, E.; Bustamante, C.; Herman, K.; Velez-Zuazo, X. Schroederichthys chilensis. In *The IUCN Red List of Threatened Species*; IUCN: Cambridge, UK, 2020; p. e.T44585A124433964. [CrossRef]
48. Carr, M.H. Effects of Macroalgal Assemblages on the Recruitment of Temperate Zone Reef Fishes. *J. Exp. Mar. Biol. Ecol.* **1989**, *126*, 59–76. [CrossRef]
49. Pérez-Matus, A.; Micheli, F.; Konar, B.; Shears, N.; Low, N.H.N.; Okamoto, D.K.; Wernberg, T.; Krumhansl, K.A.; Ling, S.D.; Kingsford, M.; et al. Kelp forests as nursery and foundational habitat for reef fishes. *Ecology* **2025**, *106*, e70007. [CrossRef]
50. Lefcheck, J.S.; Hughes, B.B.; Johnson, A.J.; Pfirrmann, B.W.; Rasher, D.B.; Smyth, A.R.; Williams, B.L.; Beck, M.W.; Orth, R.J. Are coastal habitats important nurseries? A meta-analysis. *Conserv. Lett.* **2019**, *12*, e12645. [CrossRef]

MDPI AG
Grosspeteranlage 5
4052 Basel
Switzerland
Tel.: +41 61 683 77 34

Diversity Editorial Office
E-mail: diversity@mdpi.com
www.mdpi.com/journal/diversity

www.ingramcontent.com/pod-product-compliance
Lightning Source LLC
Chambersburg PA
CBHW062036220926
51850CB00010B/2574
9783725856398